Advances in Fisheries

Advances in Fisheries

Dr. Swarnim

AMIGA PRESS. INC
Delhi -110051

AMIGA PRESS. INC
H.O :- 28-A, Indra Park, Chander Nagar
Delhi :- 110051 (India)
Tel :- 011-22504028, +91-98736-57000
Email :- amigapress@gmail.com

This edition is published by Amiga Press.Inc

ISBN : 978-81-929542-0-2

First Published 2018

PRINTED : REPLIKA PRESS PVT. LTD.

Preface

Fisheries have been known to be an important activity throughout the world, and produce more than 100 million tons of fish and fishery products and contribute to human welfare by providing a livelihood to about 200 million people, and protein supply for a billion people. However, with declining stocks as well as several evidences related to fisheries, sustainability issue became very important and has been discussed as the central topic in fishery sciences and industries. This condition is mainly encouraged by the unfortunate reality that many fisheries are in a state of crisis, and some of them demanding urgent attention. At global levels, fishing industry is a highly adaptive, market-driven and dynamically internationalized sector within the world economy. Its pressure on resources is still increasing, owing to a persistent worldwide upward trend in fish consumption, in concert with human population growth, especially in coastal zones. Global efforts are increasing and have limited the capacity of individual governments to control the pressure over fishing. This problem was associated with a variety of environmental and ecosystem problems including wastage from discards, loss of critical habitats, impact on endangered species, etc.

Fish temperature reduction is by far the most important effect of ice utilization. Therefore, the quicker the ice chills the better. Although cold-shock reactions have been reported in a few tropical species when iced, leading to a loss of yield of fillets, the advantage of quick chilling usually outweighs other considerations. The development of ad hoc fish handling methods is of course not ruled out in the case of species that could present cold-shock behaviour.

Biotechnology in fisheries and aquaculture represents a range of technologies that present opportunities to increase growth rate in farmed species, to improve nutrition of feeds for aquaculture, to improve fish health, to help restore and protect environments, to extend the range of aquatic species and to improve management and conservation of wild stocks. In this e-mail conference, the focus will be on genetic biotechnologies, with a brief treatment of related reproductive and gene banking technologies, and the appropriateness of their application in the fishery sector in developing countries. It is important to note that developing countries produce more fishery products from aquaculture, inland capture fisheries and marine capture fisheries than developed countries. The coverage of the biotechnologies here is not comprehensive, but should be enough to stimulate discussion in the conference.

Marine resource governance is a global challenge of increasing magnitude and importance. Depletion of fisheries resources, coastal habitat degradation, threats of global warming, food security at risk and rising fuel costs all contribute to the challenge. Small island developing states (SIDS) in the insular Caribbean are particularly vulnerable to natural and anthropogenic threats. Fishing and coastal communities and enterprises depend heavily on marine resources and ecosystems. Cross-scale and multi-level networks for marine resource governance may provide means for enhancing resilience and adaptive capacity in these complex social-ecological systems. Understanding how local, national and transboundary networks link actors at various levels and scales across the region in their quest to use and manage shared marine resources is key to developing policies and practices that enable good governance. The nature of such networks is being investigated through cases in small-scale fisheries governance in order to determine what features, including collaborative processes, relate to resilience and the capacity to adapt. This research is a work in progress, the conceptual framework and preliminary findings of which are shared in this paper, along with early ideas on how research outputs can be communicated and used to improve marine resource governance in the Caribbean.

The book contains all essential terms used today in the broad field of fisheries science, genetic structure, and aqiaci;tire. An invaluable reference source for researchers and teachers in fishereis and allied fields and for anyone interested in fishes.

—Editor

Contents

1

Introduction to Fishes

INTRODUCTION

Fish are an important resource worldwide, especially as food. Commercial and subsistence fishers hunt fish in wild fisheries or farm them in ponds or in cages in the ocean (see aquaculture). They are also caught by recreational fishers, kept as pets, raised by fishkeepers, and exhibited in public aquaria. Fish have had a role in culture through the ages, serving as deities, religious symbols, and as the subjects of art, books and movies. Fish are a huge group of animals, represented by over 28 000 different species. They are the oldest as well as the largest group of vertebrates. They are characterised by being almost exclusively aquatic throughout their lives. They are limbless, but have two sets of paired fins as well as a variety of single fins. They breathe predominantly by using organs known as gills. They are cold-blooded (poikilothermic), which means their body temperature varies with that of their surroundings. Their skin is usually covered with scales. Most temperate freshwater fish are called teleost (' true bone ') fish, which means that they have a bony skeleton. The only fish found in temperate freshwater that do not belong to this group are the sturgeons and the lampreys. Neither of these groups is found commonly in still-waters, although a few species of sturgeons have been introduced into still-waters in recent years. Teleost fish are extremely successful and well adapted to their environment.

Fish s pecies

There are around 50 species of fish found for at least part of their lives in the freshwaters of the British Isles. This is quite a small number of species considering the diversity of the aquatic habitats available to them and the number of species in similar waters in continental Europe. Although many of the species are native to the British Isles, there are also many that have been successfully introduced over the centuries. Some of these have settled and become well established and are now generally considered ' native ', for example the carp. Some, however, still raise concerns among conservationists about the potential impact they may have, should they become more widespread; the zander and wels catfish for example, are described as ' non-native ' and are subject to regulation. Fish species have developed and evolved to make the best use of certain habitats. Many of the species present in the British Isles are especially developed to live in rivers. These species are usually unsuited to living in still-waters as they are not adapted to this very different type of habitat. In this chapter we will concentrate on those species already adapted to the still-water, and those that are commonly stocked in still-waters with apparent success, such as the chub. The barbel has been included, as it is stocked commonly in still-waters, but it should be borne in mind that special conditions must be maintained for this species owing to its requirement for clear, oxygen-rich water. This chapter sets out some detailed information about each species, which should allow fishery managers to understand better the species under their custody. The ecology of lakes, should allow the manager to establish the best conditions for the fish species and assess any other species for likely success in any given situation.

Cyprinidae

The barbel is a large river fish with a long, streamlined, slightly flattened body. The juveniles are an overall brown colour but as they get larger they develop a dark brown back with golden bronze flanks and a creamy white underbelly. Adults generally grow to around 70 cm, but can get up to 90 cm and weigh in excess of 6 kg (the British rod-caught record 2008 is 9.837 kg). The barbel is native to the southern UK and mainland Europe. It is widespread and relatively common.

The barbel will generally mature after 3-5 years. Spawning takes place once a year in the late spring, usually May-July. Barbel spawn in small groups among gravel and stones in fast-flowing rivers. The adhesive eggs stay within the gravel bed until they hatch 10-15 days later. The fry tend to congregate in slow flowing areas where they feed on zooplankton. As they mature and get larger they tend to move out into the main flow of the river, feeding on invertebrates and even small fish in among clean gravel and associated vegetation.

The barbel is a very popular coarse fish, being one of the most sought-after river fish. They are a hard fighting fish that can grow very large, attracting specialist anglers.

Barbel have increasingly been stocked in still-waters over several years. They appear to grow and survive reasonably well, but there is little evidence that they can successfully spawn in most still-waters. It is vital that any stocking is with barbel obtained from a reputable fish farm. Barbel can sometimes be introduced from fish poached from the wild. Not only is this very harmful to the wild populations but evidence suggests that these fish find it very difficult to make the transition from rivers to still-waters. Generally barbel should not be stocked in shallow still-waters with a predominance of bottom feeders such as carp.

Common b ream (Abramis brama)

The common bream is a large, bottom-feeding species with a deep thin body. When young it is a very silvery fish, but as it matures it becomes darker and more golden brown to green. It can grow to nearly a metre in length and weigh up to 8 kg (the British rod-caught record 2008 is 8.329 kg). The common bream is native to the UK and central Europe where it is widespread and common. It is a popular food fish in eastern Europe.

Bream generally mature after 4 years. Spawning takes place in May and June in dense vegetation in shallow water. The males develop white bumps (tubercles) over the body surface and become territorial. The females spawn at night in large groups and their sticky eggs attach to the vegetation. After 5 – 10 days the eggs hatch. The fry start by feeding on zooplankton such as rotifers and then larger zooplankton such as daphnia. They quickly move on to their normal adult diet of benthic invertebrates, preferentially worms, molluscs and insect larvae such as chironomid midge larvae. The adults feed by using their extendable tubular mouth to root around in the pond bottom, taking large mouthfuls and ' winnowing ' out the food items. They are strongly shoaling fish from larval fry to late adulthood. Some of the very large individuals appear to be solitary or in small shoals. The common bream is one of the most popular of the coarse fish. It responds well to anglers ' baits and is readily caught. It grows quickly to a good size and survives well in many different conditions. As a bottom feeder it will disturb the bottom of the pond, causing suspended solids to cloud the water column. They are good in a mixed silver-fish population. The common bream should not be mixed with large numbers of carp as the species compete for the same food source.

The silver bream is a small mid-water species, with a deep, compressed body. When young it is a very silvery fish but as it matures its back becomes a darker olive/grey, although it still retains a very silvery look. Its maximum size is around 20 – 25 cm in length (the British rod-caught record 2008 is 0.425 kg). The silver bream is native to England, central and northern Europe, where it widespread and reasonably common. In England it is generally restricted to the southeast but is becoming more widespread.

Silver bream mature at around 3 – 4 years old. Spawning takes place in May – July in dense vegetation in shallow water. They spawn in groups early in the morning, scattering their sticky eggs among the submerged weeds. The eggs hatch in 4 – 8 days and begin feeding on small planktonic invertebrates, moving to larger invertebrates and plant material as they grow. Silver bream form strong shoals with individuals of a similar size and age throughout their lives. The silver bream is not a popular species for anglers, largely because it is commonly mistaken for a small common bream. Indeed it is very similar to a common bream of the same size, except the silver bream is slightly thicker in the body and has larger scales. They do make useful additions to a lake as they feed in a very different manner.

The carp is a large, bottom-feeding species with a thickset rounded body. They are usually a dark olive brown on the back, lightening to a light yellowish brown on the belly. It can grow to very large sizes, sometimes exceeding 1 m in length (the British rod-caught record 2008 is 26.9 kg). The carp is native to eastern Europe and Asia, but has now achieved worldwide distribution. It is very widespread and common in the UK, where it was first introduced in the nineteenth century.

Carp mature after 2 – 3 years. Spawning takes place in June to July in dense vegetation on the margins of the water body. The males develop small white lumps (tubercles) on the flanks and head. The females spawn at dawn and the sticky eggs attach to the weed. The eggs will hatch after 5 – 7 days and the fry start feeding on zooplankton, particularly rotifers. They quickly move to feeding on the bottom-dwelling invertebrates that make up the bulk of their diet. Their mouth is well designed for this feeding technique, being protrusible. However, carp are a very adaptable species and they can often be seen feeding on invertebrates on marginal vegetation and even off the surface. They tend to move around in small shoals of similar-sized individuals.

The carp is probably the most popular species for anglers on still-waters. It grows quickly to very large sizes and responds well to a variety of anglers ' baits. It is a very adaptable species so can survive and thrive in many different conditions. As a bottom feeder, it will disturb the sediment on the bottom of the water body, causing suspended solids to cloud the water. Carp have been bred for food over many centuries and it is one of the main global species for aquaculture in ponds. This has led to several different varieties of carp being commonly found. The most obvious of these varieties is the common carp, which is the fully scaled type. Mirror carp were bred to make it easier to remove the scales before cooking. Ornamental varieties are also common, from full-colour varieties, known as koi, to slightly gold-tinged cross breeds known as ghost carp. These are all the same species of fish, however.

The chub is a large, mid-water and surface-feeding fish with a rounded, streamlined body. When young it is a very silvery fish with an olive green back; as it gets larger its upper body darkens to a steel grey colour but retains the silvery flanks and belly. It is a large fish, attaining sizes of between 40 and 50 cm, rarely getting to up to 80 cm (the British rod-caught record 2008 is 3.9 kg). The chub is native to the UK and central Europe and is widespread and common. It is caught for food in some parts of Europe. Chub generally mature after 3 – 4 years. Spawning takes place during April – June among clean gravel and stones in flowing water. They spawn in the early morning in small groups of one female with several males in attendance. The eggs hatch within 6 – 8 days and start to feed on benthic invertebrates. As they grow older their diet changes to include larger invertebrates, fruit, vegetation and even small fish. When young, chub move around in shoals but as they mature they become much more solitary and even appear to be somewhat territorial. In rivers they migrate significant distances to feeding and spawning areas. Little is known about their activities in still-waters.

The chub is a popular fish among anglers because it is a large, hard-fighting fish that responds well to anglers ' baits. It grows quickly to a good size and appears to do very well in still-waters. Current evidence suggests that it does not successfully spawn in most still -waters, although it does spawn in large, clear-water lakes.

The crucian carp is a medium-sized, deep-bodied fish with marked lateral compression. Young fish are a golden bronze colour, which turns darker as they get older, with mature specimens being olive brown on the back through deep bronze on the flanks to a golden yellow on the belly. Adults can attain 20 – 30 cm in length and 1.5 kg in weight, although many crucian carp populations remain ' stunted ' with the mature adults rarely getting more than 15 – 20 cm and 0.5 kg (the British rod-caught record 2008 is 2.01 kg). The crucian carp is native to the UK, Europe and central Asia. There is some argument about whether it is a native to the UK and western Europe, but there seems little evidence to suggest it is introduced into these areas, where it is widespread. Crucian carp generally mature after 2 – 3 years. Spawning takes place in May – June in shallow water, in thick vegetation. The fish shed sticky eggs onto the vegetation where they take between 5 and 7 days to hatch, depending on the vegetation. Once hatched, and after the yolk sac has been used, the fry tend to form loose shoals and feed on small zooplankton. As they grow they start to feed mainly on invertebrates, such as molluscs and worms on the bottom of the pond, as well as plant material. Both the juveniles and the adults tend to live in loose shoals. They tend to occupy the mid-water and bottom of still-waters. Crucian carp prefer still-waters and particularly thrive in rich, lowland still-waters with abundant vegetation. Crucian carp have a remarkable ability to survive very low dissolved oxygen situations for long periods, leading to situations where they are often the only fish species to survive. Here they form healthy populations dominated by small ' stunted ' individuals. In larger waters they can grow much larger and seem to adopt a slightly different body shape.

This is a very popular angling species and is readily caught. It survives well in well -managed fisheries dominated by silver fish, but it is generally unsuitable for stocking in waters dominated by common carp. Crucians readily hybridise with goldfish and common carp. They are then displaced by the vigorous hybrid. They are also threatened by the non-native parasitic tapeworm *Bothriocephalus acheilognathi*.

The goldfish is a medium – sized mid-water species with a relatively deep and laterally compressed body. Wild goldfish are a deep brown colour throughout their lives, although there are many colour variants from deep orange through to almost white. Adults can get to around 30 cm and 1 kg in size. The goldfish is native to Asia, but owing to its ornamental value it has achieved worldwide distribution, being found on most continents. It is an introduced species in the UK.

Goldfish generally mature after 2 – 3 years. Spawning takes place during June and July in thick weed in the shallow margins. The eggs hatch in 4 – 7 days, depending on the water temperature, and the young feed on zooplankton. As they grow, the diet changes to small invertebrates and plant material. Goldfish tend to form loose shoals and prefer shallow, rich ponds with abundant vegetation. They are a hardy species, however, and appear to thrive in most still-water conditions.

The goldfish has become increasingly common in still-waters in the UK and Europe. It is apparently popular with some fishermen.

Goldfish readily hybridise with common carp and crucian carp, producing vigorous offspring that can successfully outcompete the original parent stock. Stocking with goldfish or hybrid strains is advised against.

The grass carp is a large mid-water feeding species with a solid, rounded body. It has a dark blue-black back, shading to silver on the belly. It can grow quite large, with adults reaching up to 100 cm in length (the British rod-caught record 2008 is 20.185 kg). Grass carp are originally from Asia, but have been widely distributed around the world as a method of weed control. They were introduced into England in the 1960s to control weed growth in some still-waters. Since then they have become widely distributed, but because of their specialised breeding requirements there appear to be no breeding populations.

Grass carp mature after 3 – 4 years. Spawning takes place in fast-flowing water in the headwaters of large rivers. Temperatures must be around 23 – 25 ° C. The eggs are unusual in that they are pelagic, being carried down river in mid-water until they hatch in the slower flows downstream. The fry start by feeding on small invertebrates, but as they get older they begin to feed on aquatic plants. Grass carp form loose shoals throughout their lives.

Grass carp are not particularly popular among anglers as they are generally difficult to catch and do not fight as hard as their size might suggest. They are not very effective in controlling weed growth as they do not feed until the waters are quite warm.

The gudgeon is a small, rounded, bottom-dwelling fish. Its back is a dark olive green to brown with lighter flanks and a silvery belly. It has dark markings along the flank and upper body. It is a very small fish, rarely reaching over 15 cm although it can grow to about 20 cm. (the British rod-caught record 2008 is 0.141 kg). Gudgeon are widely distributed throughout Europe and Asia and they are native to the UK.

Gudgeon mature at between 2 and 3 years old. They will spawn on stony or gravely bottoms, or among weed, and prefer flowing water, although they will spawn successfully in still-waters. They spawn during May or June and the eggs hatch in around 15 days. They feed on insects, molluscs, crustaceans and occasionally plant material. They are generally a solitary fish, but sometimes form into small shoals of perhaps a dozen individuals.

The gudgeon is not a popular target species for anglers, but it does do well in some still -waters where it probably makes an important contribution to the health of the lake.

The minnow is a small mid-water species with a long rounded body. It has a brown-gold body with numerous brown and black spots down the flanks usually joining to form an obvious black line along the body. It is a small fish, rarely exceeding 10 cm in length (the British rod-caught record 2008 is 0.135 kg). The minnow is a native to the UK, Europe and Asia where it is very widespread and common.

Minnows mature after 1 – 2 years. Spawning takes place in June or July, usually over clean stones or gravels. The males particularly become more colourful during spawning, with a scarlet belly and black throat. The eggs take 5 – 10 days to hatch and immediately start feeding on small invertebrates. As they grow they continue to feed on a variety of invertebrates. Minnows generally aggregate into small, loose shoals.

The minnow is not a target fish for anglers, although it can be important as a bait fish. It does well in lakes and ponds and can be a very useful addition to the lake ecology, particularly if predatory fish are present.

The orfe is a large, mid-water and surface-feeding fish with a rounded, streamlined body. When young it is a very silvery fish, but as it gets larger its upper body darkens to a brownish grey colour. Coloured ornamental varieties are common, particularly golden and blue orfe. It is a medium-sized fish, attaining sizes of between 30 and 40 cm, although it can occasionally get up to 80 cm. (the British rod-caught record 2008 is 3.7 kg). The orfe is native to central and eastern Europe and parts of Asia, where it is widespread and common. It is caught for food in some parts of Russia. It was introduced into England and is now widespread and common.

Orfe generally mature after 3 – 4 years. Spawning takes place during April and May, among weeds and clean stones in shallow water. The eggs hatch within 15 – 20 days and start to feed on benthic invertebrates. As they grow older their diet changes to include larger invertebrates, although they will take fruit, vegetation and even small fish. Orfe move around in shoals of similar-sized and aged individuals. They are found equally in rivers and still-waters.

The orfe is a popular fish among anglers because it is a good-sized, hard fighting fish that responds well to anglers ' baits. It grows quickly, to a good size and it appears to do very well in still-waters. The roach is medium-sized, mid-water fish with a relatively deep body. The young fish are silvery, but they quickly develop a dark blue – green back and light underside. Adults generally get to 20 – 40 cm and between 1 and 2 kg (the British rod-caught record 2008 is 1.899 kg). The roach is native to the UK and Europe, where it is very widespread and common throughout the lowland catchments.

Roach generally mature after 2 or 3 years. Spawning takes place once a year in the middle of spring, usually April – June. The roach spawn at dawn, shedding their sticky eggs on dense, submerged vegetation in the shallow areas of the lake. The eggs will hatch in 5 – 10 days. Once the yolk sac is used, the fry will feed on zooplankton and other small invertebrates. The juveniles prefer to live in large shoals in the shallow, marginal areas with plenty of vegetation, which provide cover and food. As they mature and get larger they move out into open water, shoaling in smaller groups of similar-sized individuals. These larger individuals feed on larger benthic invertebrates and pondweed, particularly attached algae (periphyton). They tend to occupy the mid-water to bottom of the pond. The roach is one of the most popular coarse fish in the UK. It responds well to anglers ' bait and is readily caught. It grows steadily and survives well in many different conditions. It is an excellent species in any mixed fishery as it is very adaptable. In the right conditions, it can quickly overpopulate with a very large population of small fish.

The rudd is a medium-sized, top-feeding species, with a deep, laterally compressed body. It has a dark greenish-brown back, grading through to a silvery underside. The flanks are golden brown. The iris is a deep gold with a red fleck on the upper side. The fins are normally a bright red. It can grow to 20 – 25 cm and 0.5 – 1 kg (the British rod-caught record 2008 is 2.1 kg). The rudd is native to the UK and central Europe, where it is very widespread and common.

Rudd generally mature at 2 – 4 years and live to 10 – 14 years. They spawn in late spring, usually from April to July. Rudd spawn by shedding their eggs on submerged vegetation around the margins. The eggs hatch in around 5 – 10 days and the fry stay within the protection of the vegetation, feeding on small zooplankton. As they grow they form compact shoals, feeding on the surface and in midwater. They prefer zooplankton, weed-dwelling invertebrates and terrestrial insects on the surface. Rudd appear always to spend their time in fairly compact shoals, no matter how large they get.

The rudd is a popular fish to catch, being attractive and readily taking bait, which is particularly important if other species are more reluctant to feed on anglers ' bait. It grows steadily but rarely attains large sizes. It is a very good fish for a mixed fishery, as it does not compete directly with any other fish species.

The tench is a large, bottom-feeding species, with a thickset, slightly laterally compressed body. The tench is a beautiful olive to dark green throughout its life, with small orange to red eyes. Adults generally grow slowly to a maximum length of 60 – 70 cm and can weigh up to 7 kg. (the British rod-caught record 2008 is 6.9 kg). The tench is native to the UK and Europe, although absent from the northern parts of both. The species has been widely introduced into Asia and North America.

Tench generally mature at 3 – 4 years old. Owing to their slow growth, they may only be 15 cm at this age. Spawning takes place once a year in spring, usually between May and July. They spawn among aquatic plants in shallow waters, to which their sticky eggs attach. The eggs will hatch in 4 – 5 days, and once the yolk sac is used the fry will feed on zooplank-ton, before switching to grazing among the bottom of water plants on a variety of aquatic invertebrates and attached algae. Tench prefer still-waters, or very slow flowing rivers with an abundance of macrophytes. The juveniles may form loose shoals, but as they mature they become more solitary, almost appearing to have preferred territories.

The tench is one of the most popular and easily recognisable of the coarse fish in the UK and Europe. Its popularity relates to its size and its beauty. It is generally a difficult fish to catch on rod and line. It is quite adaptable to a mixed fishery, but does not thrive in waters without good communities of aquatic macrophytes.

The pike is a large predatory fish, with a slim streamlined body and a large head with an enormous mouth full of teeth. It has a dark green, mottled back, grading through light green flanks to a creamy yellow underbelly. It is a very large fish, routinely growing to 50 – 100 cm and occasionally 140 cm and 30 kg (the British rod-caught record 2008 is 21.2 kg). The pike is widespread and abundant throughout the Northern Hemisphere in Europe, Asia and North America. It is a popular food fish wherever it is found.

Pike generally mature after 3 – 4 years, the males earlier and smaller than the females. They spawn among weed during February – May. The eggs attach to the weed and hatch after 10 – 15 days. When they hatch they start by feeding on invertebrates but quickly turn to small fish and then onto larger prey, including other vertebrates such as ducks and water rats. They are very partial to other pike and are, unsurprisingly, very solitary and territorial. It appears that females grow much larger than males, and it is said that any pike over 5 kg will be a female. They are an ambush predator, laying in wait for their prey and only striking at the very last minute when the prey is within one to two body lengths. The pike is one of the most popular freshwater fish in the UK. It grows exceptionally quickly to a large size and is a ferocious fighter, particular in the first moments as it attacks the bait.

Percidae

The ruffe is a small fish with an obvious spiny dorsal fin, large eyes and mouth. Its back is a dark green-brown colour with irregular dark blotches, whereas its underside is pale yellow to cream in colour. It usually grows to around 15 cm but can exceptionally grow to 25 cm (the British rod-caught record 2008 is 0.142 kg). The ruffe is native to much of Europe but was confined to eastern England, although its range has spread considerably in recent decades.

The ruffe matures after 1 to 2 years. They spawn in shallow water among stones and vegetation, mainly in April to May, although they probably spawn several times in a year. Initially the fry grow quickly on a diet of small invertebrates before moving on to larger invertebrates and other food such as fish eggs and young fry. The ruffe is a shoaling fish, moving around in groups throughout its life. The ruffe is not a popular fish for anglers, being such a small size and attacking baits destined for other larger target species. It has numerous alternative common names such as pope, tommy and daddie.

Perch (Perca fluviatilis)

The perch is a beautiful, deep-bodied fish with four to six dark bands over olive-green flanks. The abdomen is creamy silver with prominent red pelvic and anal fins. It can grow to 20 – 40 cm and weigh 1 – 2.5 kg (the British rod-caught record 2008 is 2.523 kg). The perch is native to the UK and Europe and is now widespread and common in the UK, northern Europe and Asia. It is a popular food fish in many countries; in England it used to be canned and sold as ' perchines ' .

Perch generally mature after 3 – 4 years and live to around 10 years. They spawn in early spring, usually March to May. Perch tend to spawn in the margins, producing long strings of eggs which lie wrapped around submerged vegetation. The eggs hatch after 10 days and the juveniles feed on small invertebrates as well as fish fry, including other perch fry. As they get older they start to feed on larger invertebrates and begin to feed actively on other fish. Small perch shoal in groups but as they get bigger they do become more solitary, with large individuals tending to move around in small shoals of just a few individuals. The perch is a popular fish for coarse fisherman, being one of the most attractive of the freshwater fish species. It responds well to anglers ' baits and is readily caught. As a predator it is a useful fish to control the populations of other fish species. However, they can mature at a very small size (10 – 15cm) and in certain circumstances they can create an abundance of small perch. It appears that perch only grow to a large size where there is a plentiful supply of small prey fish.

Zander (Sander lucioperca)

The zander is a large predatory species with a long, thick body. The back is a dark brown – green with obvious dark brown vertical stripes running to the middle of the body. It grows quite large, reaching average sizes of 50 – 70cm, occasionally much bigger to a maximum of 120 cm (the British rod-caught record 2008 is 8.7 kg). The zander is common throughout Europe. It was deliberately introduced into England (the River Great Ouse) in 1963 and has spread widely since.

Zander generally mature after 3 – 5 years. Spawning takes place from April to June in among marginal vegetation. Pairs of fish spawn together at dawn and lay sticky eggs on vegetation and stones on the bottom. Unusually, both parents will guard the eggs until they hatch 5 – 10 days after spawning. The fry start on a diet of invertebrates but move quickly on to eat mainly of other fish. The zander is a hunter, patrolling open waters and chasing down small prey fish. Occasionally it appears to hunt as a group, although it mainly hunts as an individual.

The zander is a very popular fish with anglers, being a fast-growing, large and aggressive fish. It survives well in larger, clear still-waters and may be a useful addition to a specimen water as it prefers small fish as prey. Licences are required to stock this species.

Gasterosteidae

Three-s pined s tickleback *(Gasterosteus aculeatus)*

The stickleback is one of the most common and recognisable species in freshwaters. It is a small, thin-bodied fish with an olive-brown back and silvery sides and underside. The body is covered in large bony plates and there are three obvious strong spines on the back. The stickleback is a very small fish that rarely exceeds 10 cm in length (the British rod -caught record 2008 is 7 g (0.007 kg)). Sticklebacks are widely distributed and common throughout Europe. Sticklebacks mature after 1 – 2 years. Spawning takes place in March – June. They are unusual in that the male builds a nest of fibrous plant material and attracts females to spawn within it. The male also develops a vivid red throat to help attract a mate. The eggs take 10 – 20 days to hatch and are guarded constantly by the male. This guarding behavior continues until the fry are a few weeks old. The fry feed on small invertebrates. As they get older they continue to feed on invertebrates but may extend their diet to include fish fry. Sticklebacks tend to aggregate into small shoals, only becoming solitary and territorial during spawning times.

The stickleback is not a target species for anglers. It is common in most still-waters and will almost certainly appear without help.

Nine-s pined s tickleback *(Pungitius p ungitius)*

This is a rare species. It is similar in most respects to the three-spined stickleback, except that it has 7 – 12 spines running along its back. It also goes dark with a black throat when in breeding condition.

ANGUILLIDAE

Anguillidae is a family of fishes that contains the freshwater eels. There are 19 species and 6 subspecies in this family, all in genus Anguilla. They are catadromous, meaning they spend their lives in freshwater rivers, lakes, or estuaries and return to the ocean to spawn. The young eel larvae, called leptocephali, live only in the ocean and consume small particles called marine snow. They grow larger in size, and in their next growth stage they are called glass eels. At this stage they enter estuaries and when they become pigmented they are known as elvers. Elvers travel upstream in freshwater rivers where they grow to adulthood. Some details of eel reproduction are as yet unknown, and the discovery of the spawning area of the American and European eels in the Sargasso Sea is one of the more famous anecdotes in the history of Ichthyology. The spawning areas of some other anguillid eels, such as the Japanese eel, and the giant mottled eel were also discovered recently in the western North Pacific Ocean.

European e el (Anguilla anguilla)

The eel is a medium-sized fish with a very long, cylindrical body. The anal, caudal and dorsal fins all merge to form one long fin that goes around most of the body. It has no pelvic fins. The young stages are transparent, with pigment only becoming apparent as they become elvers in the rivers. The upper surface is generally a dark green – brown with a creamy yellow belly. As they mature and get ready to migrate, the back becomes very dark with a silvery white underside. The eel can grow to very large sizes, around 1.5 m, but more commonly they are 40 – 90 cm, (the British rod-caught record 2008 is 5.0 kg). The eel is widespread and common throughout western Europe and the Mediterranean. However, dramatic declines in the number of elvers returning to the European coastline mean that there are great fears for the long-term future of this species.

Little is known about the actual spawning of eels and this mystery is only slowly being revealed..It is known that the young, leaf-like larvae of the eels drift into the waters off the European coast in early spring, where they change into glass eels before running up the rivers and developing further into elvers. Once established in freshwaters they may remain for as long as 20 years, although estimates vary from 10 years to more than 40 years. Once they decide to migrate to the spawning grounds, they brook no obstacle, even going across land to reach their goal. Once they leave our shores, evidence suggests that they journey across the Atlantic Ocean to a stable area around Bermuda and the Bahamas, known as the Sargasso Sea. Here they spawn at great depth and the adults perish. The eggs hatch and drift on the Gulf Stream back to the coast of Europe, a journey that may take as long as 3 years. Once in freshwaters, they will feed on a variety of invertebrates and small fish. They are a nocturnal and solitary species, hiding during the day and roaming at night in search of food.

The eel is a very popular species among some anglers as it grows to a large size and is very difficult to tempt to take baits. However, to some anglers it is a nuisance species as it is difficult to handle and unhook.

Siluridae

The wels catfish is a huge, bottom-dwelling scavenger, with a large, flat head and an elongated body. Its back is normally a dark, mottled dark blue/ brown black. The sides are a light reddish brown and the belly is pale. A dark mottling covers the whole body. The wels catfish can grow to a very large size, up to 3 m in length and 200 kg (the British rod-caught record 2008 is 28.1 kg). It is widespread throughout Europe, including the brackish Baltic and Black Seas. It has been introduced into several countries including England, where it is now quite widespread. Wels catfish generally mature after 4-5 years. Spawning takes place in a shallow excavation made by the male fish. The wels catfish pair up as the male attracts females into his territory. The eggs are laid in the shallow depression and then guarded by the male until they hatch. The fry start on a diet of small invertebrates but quickly move on to eat large invertebrates, fish and other vertebrates. They are mainly scavengers, using their long barbels to hunt across the bottom for food. However, they are also a keen and active predator, taking large prey items including ducks from the surface. They are a solitary species, particularly as they get larger as they are highly cannibalistic. The wels catfish is understandably popular with anglers, being such a large and strong fish. However, it is relatively difficult to catch and does not grow very large in most still-waters.

FISH ANATOMY AND PHYSIOLOGY

There are few creatures on earth that have developed such an interesting and unique set of physical characteristics as the fish. Their special adaptations have allowed them to survive in an environment completely different than humans. Water is 800 times denser than air. Water also contains less than 2% of the oxygen contained in the air. Sound and light are very distorted in the water. Yet despite these difficult obstacles in life, fish have some unique anatomical adaptions that allow them to flourish. If aquarium owners have an understanding of these basic adaptations, it will help them to better care for their fish.

Body structure

The body form of fish reflects the sort of life they lead and the habitat they live in. The basic shape is adapted to moving through water with the minimum of resistance. Water is considerably denser than air, so the ability of a fish to move freely through it is dependent on it having a streamlined shape. Unsurprisingly, therefore, most freshwater fish look very similar to each other in overall body shape. The basic shape is a torpedo, with a rounded cross section, getting smaller at the front and back, with the fins placed on the outside to allow propulsion and stability. This is known as a fusiform body shape. Different shapes have adapted from this basic pattern; for example, many still-water fish have evolved towards a slightly deeper body form because they do not need to maintain position against a current in the way a river fish would. Most other alternatives to this basic body form are adaptations for other needs, such as reproduction or feeding.

There are several ways an animal may propel itself through the water, by using combinations of fins. However, most temperate freshwater fish have evolved the successful methods of undulating their caudal fin and sometimes their trunk as the main method of propulsion.

The source of the power for this swimming method comes from the lateral body muscles lying on either side of the fish ' s backbone (around 40% of the body weight of most cyprinids). These muscles can be split into two groups: the red muscle and the white muscle. The white muscle makes up the larger mass and is usually closer in to the backbone.

The red muscle is usually found in two blocks outside the white muscle, along the midline of the fish. It is a useful generalisation to say that the white muscles are used to give fast, powerful, but brief swimming spurts, and the red muscles are used for the slow, steady, but long-lasting (sustained) swimming.

The other fins have a variety of uses, mainly to do with the stability of the swimming fish. With a fusiform body and a caudal fin powering away, there would be a tendency for the fish's head to move from side to side (yawing) and up and down (pitching) as it moves. There is also a tendency for the body of a fish to roll. These tendencies are eliminated by positioning the dorsal and anal fins on the top and bottom of the body, thus acting as keels and preventing yawing. The paired pectoral fins can be spread away from the body to prevent pitching and rolling. These paired fins are also very useful for the fish when braking and turning.

The use of the other fins decides exactly where they are placed on the body surface, depending on the lifestyle of the particular species.

There is another form of locomotion that is very important to fish, which is the slow and delicate movements that they may be required to perform. This is very dependent on one of the most important adaptations to life in water: the swim-bladder. A fish ' s body is denser than water and therefore should sink. In fact, for some groups of fish like the sharks, this is exactly what happens. However, teleost fish have developed a swim -bladder. This is a gas-filled sac, near the centre of the body, that can be filled with air. The volume of air can be increased or decreased, depending on the depth at which the fish is swimming. This gas bladder is filled with enough air to make the fish ' neutrally buoyant ' (meaning it will not sink or float). Thus a fish at rest will remain stationary in the water column. Teleost fish have developed their various fins to be able to move them to and fro very gently. This adaptation is one of the most fundamental and important aspects of the body form of fish. It allows fish to fulfil a wide variety of roles and feeding techniques and so occupy a wide range of niches in the aquatic environment.

Sensory systems

How a fish understands what is happening in the world around it is very important and quite different to how we, as humans and land animals, may understand the concept. Water is quite a different environment and requires very different senses to those needed to live on land. Eyesight particularly is probably less important in the water, whereas hearing, taste and touch are much more important. We will deal with each of the sensory systems in turn before looking at how a fish understands its surroundings by using the information it can collect with its senses.

Sight

The aquatic environment has a lower light intensity than the land. Water, and the particles suspended in it, absorbs light, so light intensity decreases quickly with depth. Fish eyes therefore have adapted to use the limited light and the alternative wavelength of light available.

The overall design of the eye is similar to that seen in most vertebrates. It is essentially a fluid-filled ball, with a small opening at one side allowing light to enter, with a lens to focus that light onto a patch of receptor cells on the other side of the ball. Teleost fish have little or no pupil control, so they cannot alter the amount of light entering the eye. Some species have a mirror-like quality to the back of the eye, formed by special cells, called the tapetum lucidum. This allows full use of all the light that does enter the eye and is a feature of nocturnal fish such as the zander.

The lens is a spherical, clear ball held in place by muscle tissue. It can be moved backwards and forwards in the eye to focus the light available. The light passes through the middle of the eye, which is filled with a liquid known as the aqueous humour. The light is focused onto the back of the eyeball, known as the retina, which is where the cells responsible for light detection are found. In fish, the retina contains two types of receptor cell, known as rods and cones, which contain the visual pigments. Most freshwater fish have cells containing all the visual pigments needed to confer full colour vision, as well as pigments that allow fish to perceive ultraviolet light.

The positioning of the eye on the fish ' s body is also important. Most fish have the eyes placed on the sides of the head. This allows good all-round vision, with some fish almost achieving the ability to look behind themselves at the same time as they look in front. However, this positioning limits the binocular vision. Binocular vision is very important for depth perception (telling how far something is away from the viewer), which is very useful for predators when they strike at their prey. So some predators, notably the pike, have eyes that are set to the front of the head to improve their binocular vision.

Smell

The nostrils are paired organs, usually found on either side of the top and front of the head. There are two openings into each of a fish's nostrils (olfactory sacs) so that water may enter through one opening and exit through the other. This flow is achieved by the design of the openings, allowing the water flow generated as the fish swims naturally to enter the front opening and exit through the rear opening. This current is further enhanced by the skin inside the nose being covered in fine hair-like structures (a layer called the ciliated epithelium), which beat in a uniform direction maintaining the flow of water. In all temperate freshwater teleost fish there is no connection between the nostril and mouth. The inside of the olfactory sac is thrown into a series of folds, serving to increase the surface area of the nose significantly. These primary folds are further folded into smaller, secondary lamellae on which can be found the chemosensory cells. There are several different forms of these cells, responsible for detecting different types of chemical. There are millions of these chemically-sensitive cells in each olfactory sac.

Tests indicate there at least four groups of chemicals that stimulate the chemoreceptor cells in the nose. These probably relate to detection of food (amino acids) and pheromones, allowing fish to detect and communicate with members of their own species and to give warning of the presence of other species.

Taste

A sense very closely related to smell and one that is well developed in fish is the sense of taste. This is the primary sense for feeding so is obviously important. Among the teleost fish, the taste buds are densely packed in the mouth; in many species they come together in a group in the roof of the mouth and develop into an organ known as the palatial organ. This organ not only tastes the food but also has touch receptors and can produce protrusions that can retain food items while non-food items are spat out.

However, taste buds are not restricted to the mouth. In some species they are commonly found on small appendages that appear around the mouth, known as barbels. Furthermore, they can even be found on other areas of the body, such as the pectoral fins and even across the general body surface.

In common with other vertebrates, fish appear to be able to distinguish the four main characteristic tastes: sweet, sour, salt and bitter. In addition to the senses of taste and smell that we would recognise as similar to those we use, some fish also have solitary chemosensory cells, similar to taste buds, scattered across the body surface. These detect a variety of chemicals and, although their purpose has not been completely established, they are likely to be used for sensing the presence of predators.

Hearing

Sound is very important underwater. Not only does sound travel five times faster in water than in air, it will also travel much further. Furthermore, sound is split into two components: particle displacement and pressure waves, both of which are important to, and can be detected by, fish using the acoustico-lateralis system. This system consists of both the inner ear in the head and the lateral line system, which is found on the trunk of the body and around the head.

The lateral line system is a long, gel-filled canal, with pores that open to the environment. The hearing system is based around a single type of mechanoreceptor cell called a hair cell. This cell is able to detect when particle displacement sound causes movement of the gel, and thus moves the hairs. The hairs are also directionally sensitive. These cells are found in both the lateral line system and the inner ear.

Using the lateral line system along the body and around the head, fish can not only detect pressure waves, but also detect the directional source of these waves. This system will also detect movement in the water. It appears to be very useful in sensing the movement of nearby fish so is functional for shoaling fish as well as predator detection. Furthermore, when a fish is swimming, it sets up movement in the water around it similar to a bow wave on a boat. The lateral line system can detect this water movement and any changes in the bow wave, for example if the fish swims close to an object.

The inner ear can also detect pressure waves using hair cells, but it has several additional functions. The inner ear consists of a series of three semicircular canals and three sac-like organs, containing small bones. These small bones, or otoliths, are held in the sac-like organs by hair cells. Being denser than the rest of the fish, they tend to ' lag behind ' if the fish moves, thus giving the fish a sense of movement. They are also affected by gravity and so give fish a sense of up and down and of balance. Owing to the difference between the density of the otoliths and that of the fish tissue, this system can also detect the particle displacement part of a sound. However, how do fish detect the important pressure wave part of sound? Strangely, this is where the swim bladder has a function.

It is suggested, and backed up by considerable evidence, that fish use the swim-bladder to convert sound pressure into particle displacement that can be transmitted to the inner ear. The swim-bladder does this by pulsating slightly in sympathy with any sound pressure waves. This pulsating creates particle displacement in the fish tissue, which the fish can ' hear ' . This works just by having the swim-bladder in reasonably close proximity to the inner ear. However, cyprinid fish, like carp, have improved on this. They have a special link of bones between their swim bladder and the inner ear called Weberian ossicles. The pulsations of the swim-bladder are transmitted directly to the inner ear along these bones. As a result, cyprinid fish have among the best hearing of any fish species.

Touch

The sense of touch is almost certainly delivered by the acoustico-lateralis system and the hair cells. These hair cells can be found on the skin, so it is likely that fish use these to sense touch.

Understanding the surroundings

In terrestrial vertebrates, vision is commonly the most important and therefore dominant sense. However, vision is more difficult in the low light intensities of the aquatic environment. Furthermore, the shifts in available light and bandwidth with increasing water depth make it even more difficult to rely on vision. For still-water fish, these problems often become even worse when the water clarity is poor, such as during an algal bloom or when there are high levels of suspended solids. Sound and water movement have very useful qualities in water that make the inner ear and lateral-line systems very important to fish. These systems not only allow them to hear sound, but also to detect very small changes in water movement, giving fish a sense of what is around them, what is moving, which direction it is in and how far away it is.

Chemical stimuli play a central role in directing many aspects of fish behaviour. This will, of course, include feeding, but also reproduction and recognition of predators, prey and members of its own species.

Fish also have a very good sense of their temporal surrounding, using various indicators to establish the time of year. For example, the pineal gland is a light-sensitive gland that is thought to play a role in timing of reproduction, growth and migration; in many species this is linked to photoperiod, which is the length of daylight during each day.

DIGESTIVE SYSTEMS

There is much variety in the possible food items available to fish. Generally fish can be categorised as herbivorous, omnivorous or carnivorous, although it is difficult to categorise many species along these lines because they have such a varied diet. Most temperate freshwater fish are omnivorous or carnivorous. There are several truly carnivorous fish, which eat fish and water mammals as well as large invertebrates. Most other fish eat some vegetation and some invertebrates.

The function of the digestive system is to hold or detain the food item, to swallow it and then break it down physically and chemically until it is small enough and soluble enough to be absorbed into the bloodstream. This is a specialised function of the fish ' s anatomy, and individual species have evolved to be very efficient in obtaining and digesting particular types of food. This can be seen clearly when comparing the digestive systems of fish feeding on different things.

Mouth

The mouth of fish is very adaptable and is quite different in different species, giving clues to the particular diet of that particular species. For all teleost fish, it is a two-part biting jaw. Some fish have teeth or bony plates for either seizing and holding prey items or crushing them. Inside the mouth is the buccal cavity. As already mentioned, this has a role in tasting the food, but it also has a role in sorting and holding food, whereas non-food items are expelled.

Towards the back of the mouth are the gill rakers, whose main function is to direct the food to the back of the mouth and prevent any going into the delicate gill tissue. This is particularly important in predatory fish where most often the prey item is still alive as it is swallowed.

Fish dentition varies with species and diet. Most fish have one or more rows of teeth in their upper and lower jaws, whose form varies according to diet. An additional set of teeth, the pharyngeal teeth, are found on the pharyngeal bones in the throat. These are used for crushing and grinding the food as it is passed back to the oesophagus. These teeth look just like molar teeth and are ground against a hard plate. They appear to use the muscle of the pectoral fins for the grinding motion.

The oesophagus of fish is very short, but expandable to allow the easy passage of large food items. Mucus is produced along the oesophagus to aid the progress of the food item and to prevent damage to the lining.

Stomach

The stomach region shows huge variation between different species, depending on their diet. This ranges from a full musculature stomach in the pike, to a slightly expanded area of a long undifferentiated tube in the carp. It can be considered as an organ of storage, mixing and beginning breakdown of food items.

In carnivorous fish, the stomach is very distinct; it is sac-like, with very elastic but muscular walls to allow for large prey items. It is cut off from the rest of the gut by a sphincter, which retains the prey in the stomach until it is sufficiently broken down. The stomach walls secrete digestive enzymes as well as acids to aid the breakdown of the prey. The muscles of the stomach wall grind the food in peristaltic waves, thus grinding and mixing the food, aiding in its breakdown from a solid into a semi-liquid ' chyme ' . This is then allowed through into the upper intestine.

In herbivorous or omnivorous fish, the stomach is much less obvious. Usually there is just a slightly expanded area of gut between the oesophagus and the first turn of the gut, which acts in a similar way to a true stomach. These species tend to concentrate on small prey items that require less work to break them down. They also tend to have well-developed pharyngeal teeth to grind up the prey before they enter the stomach.

Intestine

In the cyprinid group, and in herbivorous fish without a stomach, the intestine is a long, thin tube running backwards and forwards through the body cavity. Its function is to complete the breakdown of the food items and absorb all the nutrients released in the process.

In carnivorous fish with a stomach, the lower intestine is much shorter, but there are often projections from the main intestine known as pyloric caeca. These appear to serve a secretory function, releasing enzymes and other chemicals into the intestine to aid digestion.

For all types of fish, the pancreas is another organ that is involved in the secretion of enzymes and other chemicals, such as bicarbonates, which help digest the food and balance the pH.

Bile is produced in the liver and stored in the gall bladder, from where it is released in response to food in the upper gut. The bile also plays a role in neutralising the acidic chyme that comes down the intestine from the stomach.

There are significant differences in the anatomy and physiology of the alimentary tract of different fish species; this is in response to the normal diet of that species. It also seems that most fish are inefficient at digesting plant material, lacking any specialist adaptation to allow them to do so. In fact, there are very few true herbivores in temperate freshwater fish. Most species of fish are well adapted to feed on certain specific food items.

Respiratory and circulatory systems

Most fish obtain oxygen from the surrounding water using a respiratory organ called the gills, specially evolved for the purpose. Other body surfaces are sometimes used, at particular times during the life history, or during migrations.

Some freshwater species have adapted the ability to breathe air, allowing them to cope with low dissolved oxygen situations, or with water bodies drying up. In temperate water these are obviously of less concern and although some species, for example the crucian carp and the eel, show adaptations, most use the normal respiratory systems.

Compared with air, water holds very little oxygen and is very dense, so it takes a great deal of energy to pass the water over a respiratory surface. The gills therefore are very highly developed.

Gills

The gill structure is similar for all temperate freshwater fish. The gills consist of two sets of filaments attached to a hard gill arch, which is in turn supported by the skeleton of the branchial arch. There are four sets of these branchial arches on either side of the head. The two sets of filaments are set behind the gill arch in a V-shaped row. Each of these filaments has a row of lamellae along each surface, thereby giving an enormous surface area for gaseous exchange. The surface of these lamellae is only one cell layer thick, so there is only one cell between the circulatory system (bloodstream) and the external environment, thus reducing the diffusion distance to the minimum possible. The gills are supplied with blood direct from the ventral aorta through the afferent branchial artery. In this manner the structure provides a huge surface area with a minimum diffusion distance and an excellent blood supply for carrying the respiratory gases to and from the body.

The forward part of the branchial arch has stiff, finger-like projections, which interlock with the adjacent gill arches. These are known as the gill rakers. The primary function of these gill rakers is to prevent any food items, or other particles from getting from the mouth out across the delicate gill tissue. However, in some species they do appear also to act as a method of filtering food items out of the water flowing across the gills.

Respiration

The fish needs to achieve a steady flow of water across these respiratory surfaces so that it can breathe efficiently. There are two main methods of achieving this, the branchial pump method or ram ventilation.

As already stated, it requires a good deal of effort to force the viscous and dense water across the gills. So it would seem logical to use the forward swimming motion to achieve this. Some fish do exactly this, a method known as ram ventilation. By keeping their mouths open as they swim forward, water is forced across the gills. This method is rare for still -water fish: most use the branchial pump method for all, or for the vast majority of, the time.

The branchial pump method of respiration requires water to be actively sucked in through the mouth, and then passed across the gills before being released from the operculum. To achieve a good, steady flow of clean water requires many adaptations, and the sequence of events for each ' breath ' reveals these.

First, the mouth is opened and the buccal cavity is expanded. This draws water in through the mouth. To prevent water entering through the operculum, this is closed and a flap of skin along its edge acts as a valve, shutting this gap completely. The mouth is closed, although many fish also have a flap of skin, like a curtain, on the top of the mouth, which prevents water flowing back out of the mouth. The buccal cavity is contracted, and the opercular cavity is expanded. This forces the water out of the mouth, over the gills and into the opercular cavity. Now the opercular cavity is contracted and the operculum is opened, forcing water out. The operculum is then closed and the whole system can start again. This system is therefore run effectively by two pumps: the buccal cavity and the opercular cavity. By careful control of these two pumps, the water flow across the gills is almost constant.

The gills are important not only in the respiratory gas exchange, taking in oxygen and excreting carbon dioxide, but also in the excretion of other waste products. The gills of most teleost fish are the most important site for the excretion of nitrogenous waste products, usually in the form of ammonia and ammonium ions. Furthermore, they are important in maintaining the salt (ion) balance within the fish, which is achieved by the active uptake of salt ions like sodium (Na^+) and chloride (Cl^-).

Reproductive systems

The production of a new generation is one of the primary motivational factors for any living thing, and the various fish species have numerous adaptations to achieve this. However, in freshwater teleost fish, most species are very similar in their methods of reproduction. It is a cyclical process, usually divisible into distinct phases from maturation of one set of adults to maturation of the next generation.

Among temperate freshwater species of fish, most have very distinct breeding seasons, usually confined to 2 – 3 weeks a year. This breeding season is very closely linked to particular environmental changes that occur every year. Most teleosts have separate male and female adults. The females lay eggs, which are fertilised outside the body (oviparous). The females are very fecund (meaning they produce lots of eggs). The eggs have a relatively short, temperature-dependent incubation period. The newly hatched larvae tend to have a very short larval stage, with a very high mortality rate, before metamorphosis into recognisable fry.

Production of eggs and sperm (milt)

The testes originate as paired organs running along the body cavity. They are usually whitish or cream coloured organs that are divided into indistinct lobes and attached to the wall of the body cavity. There is a sperm duct which leads to an opening (the urinary papilla) near the rectum.

Sperm are formed from germ cells, by a process known as spermatogenesis. This involves a period of cellular multiplication, followed by cell division, a process known as meiosis, which halves the number of chromosomes. These spermatocytes then undergo a period of growth, before metamorphosing into motile sperm cells.

The ovaries are again paired organs, although in some fish, such as the perch, one of the pair of ovaries is so reduced as to be invisible to the naked eye. Ovaries are sac-like structures, usually continuous with the oviduct which leads to an opening just beside the vent. The development of the eggs is slightly more complex than the sperm. It is very similar in the initial stages, with germ cells undergoing a period of multiplication. After this there is a period of growth and then cell division, which reduces the number of chromosomes (meiosis). This meiotic division is incomplete until after the egg is fertilised, when the egg finally expels the ' lost ' half of the chromosomes. The oocytes undergo a period of significant growth, often swelling up to 100 times in size after the last cell division, as they accumulate yolk (vitellogenesis).

Most teleosts spawn more than once in their lives and the ovaries of these fish generally contain at least two distinct populations of eggs at different developmental stages. These ovaries contain a reservoir of germ cells at early stages of development and each successive annual egg production is grown from this reservoir.

Sex determination

Reproductive systems and patterns of behaviour are extremely diverse in fish and although temperate freshwater fish tend to have adopted a similar pattern, it is worth considering this variety, particularly in methods of sex determination.

Sex determination is largely under genetic control. For most fish this is, in common with most vertebrates, the XY system. In this system, those individuals with the same chromosomes (XX, homogametic) are female and those with different chromosomes (XY, heteroga-metic) are male. In other fish species, different sets of chromosomes are used to differentiate the sexes and here the male is homogametic (ZZ) and the females are heterogametic (WZ). There are also a variety of other possible genetic sex-determination systems, using multiple or even single sex chromosomes.

Although sex determination can largely be considered to be under genetic control, environmental factors can also play a role in fish. Many environmental factors, including pho-toperiod, salinity, social environment and rearing temperature have been shown to exert an influence on sex determination. The exact mechanism of this environmental control is unclear, but essentially fish that are genetically male or female can become functional females or males respectively. This change can be experimentally induced by treating juvenile fish with sex hormones (androgens or oestrogens). In the wild, it is rare to find a population of fish that are 50% male and 50% female, which may be caused by the influence of these environmental factors.

Spawning

To get the best survival for the fry, it is vital that the timing of spawning is closely linked to the cycles of availability of prey consumed by the fry. Fish are usually found to have discrete spawning seasons, timed so that the eggs hatch at a time when food items are abundant. For temperate freshwater fish, spawning is therefore usually annual and timed for various intervals in the late winter, through to early summer, depending on the fish.

The timing of spawning is therefore very important, each fish needing to be ready for those few weeks when mates will be available and the conditions correct. Egg and milt production starts not long after spawning, when there should be an abundance of food, and therefore energy, that the fish can put towards the production of gonads. The timing of the production of gonads, and the efforts to ready them for the spawning season, appear to be largely reliant on environmental factors. The two most important of these environmental factors are temperature and day length; for example, the carp will start to mature its eggs as winter breaks and spring starts. The carp uses the slowly rising water temperatures, as well as the increase in the length of daylight, to recognise the start of spring, both of these environmental cues being vital for triggering a response. These responses are controlled by the release of different hormones (mostly testosterone and oestrogen), which trigger reactions within the gonads.

These signals only get the fish so far, however, with the fish producing and maturing the eggs and milt over a time period dictated by environmental changes. However, final maturation of the eggs and milt, and actual spawning, are timed by different factors. For example, the final stages of maturation of the eggs require the production of a different hormone altogether, the maturation-inducing hormone. Environmental cues help to ensure the fish are ready to spawn at a particular time of year; however, these cues are not sensitive enough for the last part of the process. An egg, once released from the female, has to be fertilised within a very short period before it is no longer viable. So the final part of the maturation is delayed until all the fish are ready and begin courtship. For this final part of the spawning cycle, conditions must be correct, not only environmental but also physical and individual. For example, goldfish will not spawn in the absence of a suitable weed substrate to spawn on (although they will use a variety of things if no weed is available, including floating carrier bags!).

The last part of spawning is controlled by a courtship ritual, specific to an individual species, and the release of sex pheromones which allow communication between the spawning individuals. These pheromones are released into the water and sensed by the other fish participating in the spawning, which lets all the individuals synchronise their own state of readiness, so egg and milt release happens at exactly the same time. The courtship behaviour that accompanies the spawning varies from species to species, but includes visual signs such as the bright red belly of the male stickleback. Contact signals, such as spawning tubercles on male cyprinids, help to excite the female during close-contact mating rituals. Some species of teleost fish also use sound as part of their mating ritual. These sounds are commonly caused by beating muscles along the swim-bladder.

Much of the spawning ritual is to do with timing the release of eggs and milt, but it will also help in choosing the best site for laying eggs. For most still-water species, the most appropriate or best sites will be fairly limited. Laying floating eggs will mean they are susceptible to many predators. Laying sinking eggs will allow the eggs to sink to the bottom of the still-water, where the mud will almost certainly smother the eggs. So for most still -water species, laying sticky eggs among marginal vegetation is the favoured option. With most of the larger cyprinids, spawning is set for mid-spring when the phytoplankton blooms are starting, with a related bloom of zooplankton, such as rotifers, which are perfect first food for the fry. Seasonal influences have led the fish to mature, and now they start to congregate in the shallows of the still-water, around submerged or marginal vegetation. A short period of warm weather will usually bring them into spawning condition and the females will start to dart through the vegetation, usually first thing in the morning. The males, which have already formed large white lumps on the head and flanks (tubercules), will dash after each female, nudging her and rubbing alongside. Usually three or four males will be following each female. This activity may go on for several hours each day, for a couple of days. Eventually, when all the fish are ready, the female will again race through the weed, shedding eggs and pheromones as she goes. Several males will race alongside her, shedding milt.

Once released, the eggs and milt come together and fertilisation occurs. At this point most of the adults of teleost fish species will have no further influence on the eggs, and there is no parental care. In fact, some cyprinids will begin eating eggs soon after spawning. An exception is the stickleback male, which builds a nest and, once the female has spawned within it, will guard the eggs until they hatch.

DEVELOPMENT OF EGGS AND LARVAE

At fertilisation the egg recombines the two sets of chromosomes found in the milt and egg. The eggs and milt were formed in a process (meiosis) that halved the number of chromosomes, so fertilisation reinstates the full number of chromosomes, half from the mother and half from the father. This fertilised, single cell starts to divide quickly, forming a noticeable bulge of cells on one side of the yolk sac. Eventually this group of cells start to differentiate and form a band around the yolk sac. As this process continues a noticeable head starts to form, and features such as eyes start to be recognisable. The time it takes for hatching of the fish is very temperature dependent, but is usually between 60 and 120 degree days (2 days at 18°C would be 2 × 18 = 36 degree days). When the fish first hatch, they will not be fully developed. The yolk sac will feed the embryo for a further few days as the larval fish continues to develop. Most freshwater fish spend very little time as larvae and are very vulnerable at this stage. They metamorphose quickly into little fish, commonly known as fry, and start to feed on their normal diet. The obvious exception to this is the common eel, which may spend as long as 3 years as larvae.

Coordination and control systems

In complex animals like fish, the control and coordination of activities is achieved either by electrical impulses in the nervous system, or by chemicals (hormones) being released into the bloodstream (the endocrine system). Although the two systems are often dealt with differently, they are essentially very similar, and closely linked. In simple terms, the signals, whether chemical or electrical, serve to regulate the function of the organs within the fish's body. Signals are sent from the ' control centre ' to the effector organ, which responds with a particular set of actions. Information about these actions is then transmitted back to the control centre which may then further modify the actions or stop sending signals. The major difference between the nervous system and the endocrine system is one of speed: the nervous system gives rapid control over the body functions; the endocrine system gives slow control.

Nervous systems

The most anterior parts of a fish's brain are the olfactory bulbs. These are connected to the two lobes of the cerebrum by stalks. In fish the cerebrum is primarily involved with the sense of smell. It also seems to control behaviors such as taking care of the young and exploring the environment. The optic lobes process information from the eyes. The cerebellum coordinates body movements and the medulla controls internal organ functions and maintains balance. Posterior to the brain is the spinal cord, which is the hollow dorsal nerve cord that characterizes chordates. The spinal cord is enclosed and protected by the vertebral column. Between each set of vertebrae, a pair of spinal nerves exits the cord and connects to the internal organs and muscles. Most fish have superbly designed sense organs. Fish like trout and salmon, which are active in the daylight, have well-developed eyes and color vision nearly equal to the average human. They also have extraordinary senses of taste and smell. Chemo receptors (chemo-means chemical) are located all over the head and much of the body surface. Salmon can distinguish between the odor of their own home stream and that of another stream while they are still far out at sea. Most fish have ears inside their heads, but they do not hear well. However, a series of pores connected to canals beneath the skin cover the head and the sides of their body. This system, called the lateral line system, detects the motion. Some fish-electric eels, catfish, and sharks-can detect electricity.

The nervous system of fish consists of the brain and spinal cord, known as the central nervous system (CNS). It is based upon electrical activity in the basic cells of the system, the neurons. There are two major components of the nervous system, the voluntary and the automatic systems. These can be looked at as impulses that the fish has some control over (voluntary), and the automatic impulses that the fish has no control over. So, for example, hunting for prey, particularly the final snatch, is all controlled by the voluntary nervous system, whereas the beating of the heart or the movements in the stomach muscles will all be controlled by the autonomic (automatic) nervous system.

Endocrine systems

The endocrine system is made up of the various organs (glands) that secrete hormones in response to certain stimuli. These hormones will then control certain aspects of the fish ' s behaviour, or the development of certain organs such as the gonads. For example, the pituitary organ is at the base and rear of the brain. It is the most complex of the endocrine glands and controls the function of several other endocrine glands. Its most important functions relate to the release of gonatrophic hormones, which stimulate the growth and development of the gonads, and the release of growth hormones which stimulate growth.

Immune systems

All fish are subject to a wide range of possibly dangerous parasites, bacteria and viruses, collectively known as pathogens or disease-causing organisms. To protect themselves, fish have developed a very effective defence system, their immune system. This system can be split into two separate systems, the non-specific and the specific.

Specific immune system

This deals with specific pathogens that have established themselves on or in the fish. It is the basis for ' acquired immunity ' , as the fish can ' learn ' to deal with specific pathogens and use this information to deal more effectively with the same pathogen later in life. This can be passed on to the next generation. It is also the basis for vaccinations. Every organism has a unique set of proteins and chemicals that make up its body. Any invading pathogen will have slightly different proteins on its outer surface. These proteins will be unlike anything found normally within the body of the fish. These proteins are known as antigens. The fish can recognise these proteins as being alien or ' non-self ' , and they react by producing antibodies. These antibodies are also proteins and react to the presence of the antigens by attaching to them. Once attached, the antibody can act in one of several ways: it can attract white blood cells, particularly the phagocytes, which destroy the pathogen, it can prevent the pathogen from reproducing or it can prevent the pathogen from releasing toxins. This system is very effective, but it may take several weeks for the fish to develop sufficient antibodies to deal with a rapidly reproducing pathogen. During this time, the fish may succumb to the pathogen. However, the real benefit of the specific immune system is only fully seen when the fish is attacked by a pathogen for the second time. If the fish survives and recovers from the first infection it develops a ' memory ' of the antigens from that particular pathogen. If it is subjected to another attack from the same pathogen it can now use the memory cells to start producing antibodies straight away, thus eliminating the pathogen before it has a chance to get out of control. This is known as acquired immunity.

Osmoregulatory control system

Fish, and other aquatic organisms, have a particular problem relating to maintaining the amount of water in their bodies. This is due to a process known as osmosis, which is diffusion of water through a semi-permeable membrane, from a dilute to a concentrated solution. This means that fish living in freshwater are constantly taking water in across their gills and other membranes. This is because they will have body fluids that are more concentrated than the external freshwater environment. This influx of water must be dealt with, otherwise the fish will quickly die as their body fluids become too dilute.

The primary means of ridding the body of excess water is by excreting copious amounts of dilute urine, produced in the kidneys. However, this can lead to further problems. No matter how effective the kidneys are at removing the essential salts (e.g. chloride, Cl-, and sodium, Na^+) from the urine before it is expelled, the fish can still lose too much salt. To maintain salt balance, fish have developed a salt uptake mechanism in the gills, which is highly efficient. Fish will also gain salts in their normal diet.

In short, freshwater fish will take in water across its gills and other membranes. To rid the body of excess water the kidneys have adapted to produce large quantities of dilute urine. Ammonia secretory activities are also undertaken by the gills, which use the secretion of ammonium ions to aid the active uptake of salt ions, such as sodium and chloride, which are lost during the production of the copious urine. The regulation of the salt and water balance is an important role for the endocrine system. The salt – water balance is monitored and regulated by the endocrine system and hormones, for example prolactin, which is secreted to regulate activities in the body such as the amount of water removed by the kidneys and the level of sodium uptake by the gills.

2

Advanced Methods of Fresh Fish Handling

BASICS OF FRESH FISH HANDLING AND USE OF ICE

Throughout history, man has preferred to consume fresh fish rather than other types of fish products. However, fish spoil very quickly and man has had to develop methods to preserve fish very early in history.

Keeping and transporting live fish

The first obvious way of avoiding spoilage and loss of quality is to keep caught fish alive until consumption. Handling of live fish for trade and consumption has been practised in China with carp probably for more than three thousand years. Today, keeping fish alive for consumption is a common fish-handling practice both in developed and developing countries and at both artisanal and industrial level. In the case of live fish handling, fish are first conditioned in a container with clean water, while the damaged, sick and dead fish are removed. Fish are put to starve and, if possible, water temperature is reduced in order to reduce metabolic rates and make fish less active. Low metabolic rates decrease the fouling of water with ammonia, nitrite and carbon dioxide that are toxic to fish and impair their ability to extract oxygen from water. Such toxic substances will tend to increase mortality rates. Less active fish allow for an increase in the packing density of fish in the container.

A large number of fish species are usually kept alive in holding basins, floating cages, wells and fish yards. Holding basins, normally associated with fish culture companies, can be equipped with oxygen control, water filtering and circulation and temperature control. However, more simple methods are also used in practice, for instance large palm woven baskets acting as floating cages in rivers (China), or simple fish yards constructed in a backwater of a river or rivulet for large "surubi" (Platystoma spp.), "pacu" (Colossoma spp.)

and "pirarucu" (Arapalma gigas) in the Amazonian and Parana basins in South America.

Methods of transporting live fish range from very sophisticated systems installed on trucks that regulate temperature, filter and recycle water and add oxygen (Schoemaker, 1991), to very simple artisanal systems of transporting fish in plastic bags with an oxygen supersaturated atmosphere (Berka, 1986). There are trucks that can transport up to 50 t of live salmon; however, there is also the possibility of transporting a few kilo-grammes of live fish relatively easily in a plastic bag.

By now a large number of species, inter alia, salmon, trout, carp, eel, seabream, flounder, turbot, catfish, Clarias, tilapias, mussels, oysters, cockles, shrimp, crab and lobster are kept alive and transported, very often from one country to another.

There are wide differences in the behaviour and resistance of the various species. Therefore the method of keeping and transporting live fish should be tailored according to the particular species and the length of time it needs to be kept outside its natural habitat before slaughtering. For instance, the lungfish (Protopterus spp.) can be transported and kept alive out of water for long periods, merely by keeping its skin moist.

Some species of fish, noticeably freshwater fish, are more resistant than others to changes in oxygen in solution and the presence of toxic substances. This is probably due to the fact that their biology is adapted to the wide yearly variations in water composition presented by some rivers (cycles of matter in suspension and dissolved oxygen). In these cases, live fish are kept and transported just by changing the water from time to time in the transport containers. This method is widely used in the Amazonian, Parana and Orinoco basins in South America; in Asia (particularly in the People's Republic of China, where also more sophisticated methods are used) and in Africa (N'Goma, 1993).

The most recent development is the keeping and transporting of fish in a state of hibernation. In this method, the body temperature of live fish is reduced drastically in order to reduce fish metabolism and to eliminate fish movement completely. The method greatly reduces death rates and increases package density, but careful temperature control should be exercised to maintain the hibernation temperature. There is an appropriate hibernation temperature for each species. Although the method is already utilized for instance to transport live "kuruma" shrimp *(Penaeus japonicus)* and lobster in pre-chilled wet sawdust, it should be considered an experimental technique for most of the species. Although keeping and transporting live fish is becoming more and more important, it is not a viable solution for most of the bulk fish captures in the world.

Chilling fish with ice

Historical evidence proves that the Ancient Chinese utilized natural ice to preserve fish more than three thousand years ago. Natural ice mixed with seaweed was also used by the Ancient Romans to keep fish fresh.

However, it was the development of mechanical refrigeration which made ice readily available for use in fish preservation.

In developed countries, particularly in USA and some European countries, the tradition of chilling fish with ice dates back more than a century. The practical advantages of utilizing ice in fresh fish handling are therefore well established. However, it is worthwhile for young generations of fish technologists and newcomers to the field, to review them, paying attention to the main points of this technique. Ice is utilized in fish preservation for one or more of the following reasons:

Temperature reduction. By reducing temperature to about 0°C the growth of spoilage and pathogenic micro-organisms is reduced, thus reducing the spoilage rate and reducing or eliminating some safety risks. Temperature reduction also reduces the rate of enzymatic reactions, in particular those linked to early *post mortem* changes extending, if properly applied, the *rigor mortis* period.

Fish temperature reduction is by far the most important effect of ice utilization. Therefore, the quicker the ice chills the better. Although cold-shock reactions have been reported in a few tropical species when iced, leading to a loss of yield of fillets, the advantage of quick chilling usually outweighs other considerations. The development of ad hoc fish handling methods is of course not ruled out in the case of species that could present cold-shock behaviour.

Melting ice keeps fish moist. This action mainly prevents surface dehydration and reduces weight losses. Melting water also increases the heat transport between fish and ice surfaces (water conducts heat better than air): the quickest practical chilling rate is obtained in a slurry of water and ice (e.g., the CSW system). If, for some reason, ice is not utilized immediately after catching the fish, it is worthwhile keeping the fish moist. Evaporative cooling usually reduces the surface temperature of fish below the optimum growth temperature of common spoilage and pathogenic bacteria; although it does not prevent spoiling. Ice should also be utilized in relation with chilling rooms to keep fish moist. It is advisable to keep chilling room temperature slightly above 0°C (e.g., 3-4°C).

However, water has a leaching effect and may drain away colour pigments from fish skin and gills. Ice melting water can also leach micronutrients in the case of fillets and extract relatively large amounts of soluble substances in some species (e.g., squid).

Depending on the species, severity of leaching and market requirements, an ad *hoc* handling procedure may be justified. In general, it has been found that drainage of ice meltwater is advisable in boxes and containers and that permanence of fish in chilled sea water (CSW) and refrigerated seawater (RSW) should be carefully assessed if leaching and other effects (e.g., uptake of salt from the seawater, whitening of fish eyes and gills) are to be avoided. During the past there was much discussion about allowing drainage from one fish box to another, and consequent reduction or increase of bacterial load by washing with drainage water. Today, apart from the fact that in many cases box design allows for external drainage of each box in a stack, it is recognized that these aspects have less importance when compared with the need for quick reduction in temperature.

Advantageous physical properties. Ice has some advantages when compared with other cooling methods, including refrigeration by air. The properties can be listed as follows:

(a) *Ice has a large cooling capacity.* The latent heat of fusion of ice is about 80 kcal/kg. This means that a comparatively small amount of ice will be needed to cool 1 kg of fish.

For example, for I kg of lean fish at 25°C, about 0.25 kg of melted ice will be needed to reduce its temperature to 0°C. The reason why more ice is needed in practice is mainly because ice melting should compensate for thermal losses.

The correct understanding of this ice characteristic is the main reason for the introduction of insulated fish containers in fish handling, particularly in tropical climates. The rationale is: ice keeps fish and the insulated container keeps ice. The possibility to handle fish with reduced amounts of ice improves the efficiency and economics of fresh fish handling (more volume available for fish in containers, trucks and cold storage rooms, less weight to transport and handle, reduction in ice consumption, less water consumed and less water drained).

(b) *Ice melting is a self-contained temperature control system.* Ice melting is a change in the physical state of ice (from solid to liquid), and in current conditions it occurs at a constant temperature (0°C).

This is a very fortunate property without which it would be impossible to put fresh fish of uniform quality on the market. Ice that melts around a fish has this property on all contact points. In the case of mechanical refrigeration systems (e.g., air and RSW) a mechanical or electronic control system (properly tuned) is needed; nevertheless, controlled temperature will be always an average temperature.

Depending on the volume, design and control scheme of mechanical refrigeration systems, different temperature gradients may appear in chill storage rooms and RSW holds, with fish slow freezing in one comer and maybe above 4°C in another comer. Although the need for proper records and control of temperature of chill storage rooms has been emphasized recently in connection with the application of HACCP (Hazard Analysis Critical Control Point) to fresh fish handling, it is clear that the only system that can assure accurate temperature control at the local level (e.g., in any box within a chill storage room) is ice melting.

Ice made of sea water melts at a lower temperature than fresh water ice, depending on the salt content. Theoretically with 3.5 % of salt content (the average salt content of seawater) seawater ice will melt at about-2.1°C However, as ice made out of seawater is physically unstable (ice will tend to separate from salt), brine will leach out during storage lowering the overall temperature (and this is the reason why sea water ice always seems wet). In these conditions, fish may become partially frozen in storage conditions and there may be some intake of salt by the fish muscle. Therefore, it cannot be said that ice made out of seawater has a proper self-controlled temperature system.

There is a narrow range of temperature below 0°C before fish muscle starts to freeze. The freezing point of fish muscle depends on the concentration of different solutes in the tissue fluids: for cod and haddock, it is in the range of-0.8 to-1 °C, for halibut-1 to-1.2°C, and for herring about-1.4°C (Sikorski, 1990).

The process of keeping fish below 0°C and above the freezing point is called superchilling, and it allows achievement of dramatic increases in overall keeping times. In principle it could be obtained using seawater ice or mixtures of seawater and freshwater ice, or ice made out of a 2% brine and/or mechanical refrigeration. However, in large volumes it is very difficult to control temperature so precisely and temperature gradients, partial freezing of fish in some pockets and hence lack of uniformity in quality are unavoidable.

Convenience. Ice has a number of practical properties that makes its use advantageous. They are:

(a) It is *a portable cooling method.* It can be easily stored, transported and used. Depending on the type of ice, it can be distributed uniformly around fish.

(b) *Raw material to produce ice is widely available.* Although clean, pure water is becoming increasingly difficult to find, it is still possible to consider it a widely available raw material. When there is no assurance that freshwater to produce ice will be up to the standard of drinking water, it should be properly treated, e.g., chlorination.

Clean seawater can also be utilized to produce ice. Ice from seawater is usually produced where freshwater is expensive or in short supply. However, it should be remembered that harbour waters are hardly suitable for this purpose.

(c) *Ice can be a relatively cheap method of preserving fish.* This is particularly true if ice is properly produced (avoiding wastage of energy at ice plant level), stored (to avoid losses) and utilized properly (not wasted).

(d) *Ice is a safe food-grade substance. If* produced properly and utilizing drinking water, ice is a safe food substance and does not entail any harm either to consumers or those handling it. Ice should be handled as food.

Extended shelf life. The overall reason for icing fish is to extend fresh fish shelf life in a relative simple way as compared to storage of un-iced fish at ambient temperatures above 0°C. However, extension of shelf life is not an end in itself, it is a means for producing safe fresh fish of acceptable quality. Most landed fish can be considered a commodity, that is, an article of trade. Unlike other food commodities, it is usually highly perishable and it is thus in the interest of the seller and the buyer to ensure fish safety at least until it is consumed or further processed into a less perishable product. Ice and refrigeration in general, by making possible extension of fish shelf life, convert fresh fish into a true trade commodity, both at local and international level.

Ice is used to make fish safe and of better quality to consumers. It is also used because otherwise the current fish trade at local and international level would be impossible. Shelf life is extended because there is a strong economic reason to do so. Fishermen and fish processors who fail to handle fresh fish appropriately ignore the essence of their business. The inability to recognize fresh fish also as a trade commodity is at the root of misunderstandings and difficulties linked to the improvement of fish handling methods and prevention of post-harvest losses.

Types of ice

Ice can be produced in different shapes; the most commonly utilized in fish utilization are flake, plate, tube and block. Block ice is ground before being utilized to chill fish. Ice from freshwater, of whatever source, is always ice and small differences in salt content or water hardness do not have any practical influence, even if compared with ice made out of distilled water. Cooling capacity is expressed by weight of ice (80 kcal/kg); the same volume of two different types of ice will not have the same cooling capacity. Ice volume per unit of weight can be more than twice that of water, and this is important when ice stowage and volume occupied by ice in a box or container are considered. Ice necessary to cool fish to 0°C or to compensate for thermal losses is always expressed in kilogrammes.

Under tropical conditions ice starts to melt very quickly. Part of the melted water drains away but part is retained on the ice surface. The larger the ice surface per unit of weight the larger the amount of water retained on the ice surface. Direct calorimetric determinations show that at 27°C the water on the surface of flake ice at steady conditions is around 12-16% of the total weight and in crushed ice, 10-14%. To avoid this problem, ice may be subcooled; however, under tropical conditions this effect is quickly lost. Therefore a given weight of wet ice will not have the same cooling capacity as the same weight of dry (or subcooled) ice, and this should be taken into account when making estimations of ice consumption.

There is always the question of which is the "best" ice to chill fish. There is no single answer. In general, flake ice will allow for an easier, more uniform and gentle distribution of ice around fish and in the box or container and will produce very little or no mechanical damage to fish and will chill fish rather more quickly than the other types of ice. On the other hand, flake ice will tend to occupy more volume of the box or container for the same cooling capacity and if wet, its cooling capacity will be reduced more than the other types of ice (since it has a higher area per unit of weight).

With crushed ice there is always the risk of large and sharp pieces of ice that can damage fish physically. However, crushed ice usually contains fines that melt quickly on the fish surface and large pieces of ice that tend to last longer and compensate for thermal losses. Block ice requires less stowage volume for transport, melts slowly, and contains less water at the time it is crushed than flake or plate ice. For these reasons, many artisanal fishermen utilize block ice (e.g., in Colombia, Senegal and the Philippines).

Probably tube ice and crushed ice are more suitable for use in CSW systems if ice is wet (as it normally is under tropical conditions), since they will contain less water on their surfaces. There are also economic and maintenance aspects that may play a role in deciding for one type of ice or another. The fish technologist should be prepared to analyze the different aspects involved.

Cooling rates

Cooling rates depend mainly on the surface per unit of weight of fish exposed to ice or chilled ice/water slurry. The larger the area per unit of weight the quicker the cooling rate and the shorter the time required to reach a temperature around 0°C at the thermal centre of the fish. This concept is also expressed as "the thicker the fish the lower the cooling rate".

Small species such as shrimp, sardines, anchovies and jack mackerels cool very quickly if properly handled (e.g., in CSW or CW). Large fish (e.g., tuna, bonito, large sharks) could take considerable time to cool. Fish with fat layers and thick skin will take longer to cool than lean fish and fish with thin skin of the same size.

In the case of large fish, it is advisable to gut them and to put ice into the empty belly as well as around it. In large sharks, gutting alone may not be enough to prevent spoilage during chilling, and therefore it is advisable to gut the shark, to skin it and to cut the flesh into sizeable portions (e.g., 2-3 cm thick) and to chill them as soon as possible. Chilled sea water (CSW) has in this case the advantage of extracting some of the urea present in shark muscle. However, this is an extreme case, since in current situations fillets kept in ice will last less time than gutted fish or whole fish (because of the unavoidable microbial invasion of the flesh) and will lose soluble substances.

Typical curves for cooling fish in ice, using different types of ice and chilled water (CW).

The quickest method to chill fish is with chilled water (CW) or chilled sea water (CSW), although the practical difference with flake ice is not great. There are, however, noticeable differences after the quick initial drop in temperature with crushed block ice and tube ice, due to differences in contact areas between fish and ice and flow of melt-water.

Cooling curves may also be affected by the type of container and external temperature. Since ice will melt to cool fish and simultaneously to compensate for thermal losses, temperature gradients may appear in actual boxes and containers. This type of temperature gradient could affect the cooling rate, particularly in boxes at the top or side of the stacks, and more likely with tube and block crushed ice.

Curves such as those are useful to determine the critical limit of chilling rates when applying HACCP to fresh fish handling. For instance, in specifying a critical limit for chilling fish "to be at 4.5°C in the thermal centre in no more than 4 hours", in the case could be achieved only by using flake ice or CW (or CSW).

In most cases the delay in reaching 0°C in the thermal centre of the fish may not have much practical influence because the surface temperature of the fish will be at 0°C. On the other hand, warming-up of the fish is much riskier because the fish surface temperature (which is actually the riskiest point) will almost immediately be at the external temperature, and therefore ready for spoilage. As large fish will take longer than small fish to warm up and also have less surface area (where spoilage starts) per unit of volume than small fish, they usually take a little longer to spoil than small fish. This circumstance has been widely used (and abused) in practice in the handling of large species (e.g., tuna and Nile perch).

Small species will warm up very quickly and definitely more quickly than large species (warming-up the same reason for which they cool faster). Although warming-up studies of fresh fish have received little attention in the past, they are necessary within an HACCP scheme, to determine critical limits (e.g., maximum time fish can be handled without ice in a fish processing line).

With application of HACCP and HACCP-based systems, thermometers including electronic thermometers, should be a standard tool in fish processing plants. Therefore, it is advisable to perform fish cooling and warming-up trials on actual conditions.

Fatty fish have lower cpf values than lean fish and, in theory, require less ice per kilogramme than lean fish; however, for safety purposes it is advisable to make calculations as if fish were always lean. Refinements in the determination of cpf are possible; however, they do not drastically alter the results. The theoretical quantity necessary to cool fish to 0°C is relatively small and in practice much more ice is used to keep chilled fish. If we relate the proper fish handling principle of surrounding middle and large sized fish with ice, to the approximate dimensions of ice pieces, it is clear that with some types of ice (tube, crushed block and plate) greater quantities are required for physical considerations alone.

However, the main reason for using more ice is losses. There are losses due to wet ice and ice spilt during fish handling, but by far the most important losses are thermal losses.

It is clear that there is little practical possibility in tropical countries to handle chilled fish in plastic boxes exposed to the sun. An increase in ice consumption, even if less dramatic than in plastic boxes, can be measured in insulated containers exposed to the sun. The obvious advice in this case is to keep and handle fish boxes and containers in the shade. This measure can be complemented by covering the boxes or containers with a wet tarpaulin. The wet tarpaulin will reduce the temperature of the air in contact with boxes and containers to the wet bulb temperature (some degrees below the dry bulb temperature, depending on the Equilibrium Relative Humidity-ERH-of the air), and will practically stop noticeable radiation effect (since there are always radiation effects between a body and its background).

Ice consumption in stacks of boxes and containers

In a stack of boxes or containers not all of them will lose ice in the same way. The results of an ice meltage test conducted on a stack of boxes. Boxes or containers at the top will consume more ice than boxes and containers at the bottom, and those in the middle will consume less than either.

Jensen and Hansen (1973) and Hansen (1981) presented a system ("Icibox"), mainly for artisanal fisheries. In this system, a stack of plastic boxes were insulated by placing wooden frames, filled with polystyrene, at the top and at the bottom of the stack, and covering the whole with a case made out of canvas or oil skin. A similar system, composed of stacks of styropor boxes, accommodated in a pallet, and covered by an insulated mat of high reflective

(Al) surface, is used in practice for shipment of fresh fish by air (e.g., it is utilized to ship fresh fillets of Nile perch from Lake Victoria to Europe). The use of chill rooms drastically reduces the ice consumption in plastic boxes, avoiding the need of re-icing. In a fish handling system chilling fish with ice, mechanical refrigeration is used to reduce the ice consumption and not to chill fish.

Although analytical models of ice consumption can be applied directly to estimate the ice consumption in simple and repetitive fish handling operations, their main importance is that they can help in arriving at solutions for the proper handling of chilled fish in rational. way.

Ice consumption in the sides of boxes and containers

It is necessary to bear in mind that ice will not melt uniformly in the interior of a box or container, but meltage will follow the pattern of temperature gradients between the interior of the box/container and the ambient. It is clear that the bottom and top of boxes and containers should receive more ice to compensate for thermal losses, the top receiving more ice than the bottom. However, in practice more ice should also be put in the sides of boxes and containers.

It can be seen that ice is still abundant on topof the box. However, after a period of storage in a chill room, ice has melted, mainly on the sides, leaving some fish and parts of fish exposed to the air with a consequent rise in temperature and dehydration. In addition, ice and fish have formed a compact mass that can produce physical damage to exposed fish when the box is moved.

In chilled fish onboard fishing vessels or transported by truck, this problem may not exist if there is a continuous gentle movement which allows for ice melt water from the top to move to the sides. However, in chill rooms or storage rooms (insulated containers) it would be advisable to re-ice if this problem is observed. Under tropical conditions this effect is observed, even with insulated containers, in less than 24 hours of storage.

NEW ADVANCES

Three recent scientific advances are providing good news to doctors and patients dealing with bladder cancer. The first confirms that a novel, noninvasive test can detect bladder cancer with unprecedented accuracy. The second reveals the discovery of the "bladder-cancer gene," which is expected to lead to more effective, individualized treatments of the disease, particularly in its recurrent form. The third discusses a connection made between a specific oncogene and the development of bladder cancer. All three discoveries are poised to provide laboratories with better tools to help doctors and their patients make the best possible treatment decisions.

FISH technology

In the first study, reported in the February 2004 journal Urology, researchers found that utilizing fluorescence in situ hybridization (FISH) technology when testing urine specimens for bladder-cancer markers was 92% effective in detecting those markers, compared with a sensitivity of only 64% for traditional cytology screenings.

FISH technology works on voided urine by determining the absence or corruption of certain chromosomes in it—namely numbers 7 and 9, among others. Numerous scientific studies over the past decade have linked the deletion or corruption of these two chromosomes to a high likelihood of bladder cancer.

Through FISH analysis, laboratories will be able to provide initial analysis more quickly, and with a higher degree of accuracy, to physicians, so that they and their patients can make an informed biopsy decision and begin designing potential treatment protocols much sooner. Not surprisingly, recent therapeutic studies have concluded that early diagnosis, coupled with a combined treatment regimen, leads to the highest survival rate among bladder-cancer patients.

The "missing link"

In the second study, reported in the April 2004 issue of the journal Oncogene, scientists at the British Institute for Cancer Research identified E2F3, a gene they call "the missing link in our understanding" of bladder cancer. Researchers found a direct correlation between the overexpression of the gene and the proliferation of bladder-cancer cells.

Additionally, they noted that the amount of E2F3 proteins present was a significant indicator of the seriousness of the cancer. Higher amounts were associated with larger tumors, while smaller amounts indicated smaller, Stage I malignancies. The study authors expect their work will lead to more targeted treatment and better disease-progression prediction in individual patients.

Oncogene CDC91L1

The third discovery, made by a group of scientists from Johns Hopkins University Medical Center in Baltimore and the National Institutes of Health in Bethesda, MD, and reported in the April 2004 issue of the journal Nature Medicine, is of an oncogene called CDC91L1. Researchers theorize that the overexpression of this gene is a major culprit in the development of primary bladder cancer. Relying on the knowledge that cancer begins by an activation or amplification of these oncogenes, the scientists traced what is known as a germline translocation, or break, in the DNA of a bladder-cancer patient, and found that the overexpression of CDC91L1 was the only consequence of that break. Researchers confirmed their theory through a subsequent in vivo study using 61 nude mice.

PROMISING ADVANCES

These genetic discoveries are perhaps the most promising in the quest for faster, more accurate laboratory diagnosis, which can then lead to more successful treatment of disease. Moreover, these recent advances are likely to spur more interest in further research, promoting a higher level of understanding among clinicians about the development and progression of bladder cancer and other malignancies.

Such further advances are clearly critical, as the sobering facts about this often-overlooked disease continue to be confirmed and expanded by researchers. Bladder cancer is the ninth most common form of cancer in the world today, accounting for 330,000 new cases—and 100,000 deaths—each year. Bladder cancer is also one of the most chronic cancers, recurring in almost 70% of patients.

Preventive Strategies

On the preventive front, previous research identifying smoking as a chief cause of the disease has been corroborated in numerous medical journal reports over the past several years. The International Agency for Research on Cancer (IARC) estimates that 65% of male bladder-cancer patients and 30% of female bladder-cancer patients are smokers. In fact, two studies—one in the Canadian Journal of Urology and one in the Journal of the National Cancer Institute—found that smokers are more than twice as likely to develop bladder cancer than nonsmokers, and that the rate at which female smokers develop bladder cancer is "significantly higher" than the rate of male smokers.

Clearly, bladder cancer is a disease that is troubling in its prevalence, yet the advances in our understanding of the disease bode well for patients and clinicians alike. From our knowledge about preventive strategies, to our ability to detect the disease earlier, easier, and more accurately than ever, to the prospect of new, effective gene-therapy treatments, to the prospect of even more advancements in research and development, there are many reasons to be optimistic about the future for those affected by bladder cancer.

FISH HANDLING IN ARTISANAL FISHERIES

Artisanal fisheries, existing both in developed and developing countries, encompass a very wide range of fishing boats from pirogues and canoes (large and small) to small outboard and onboard engine vessels, utilizing also a variety of fishing gears. It is difficult to find a common denominator; however, from a fish handling point of view, artisanal vessels handle relatively small amounts of fish (when compared with industrial vessels) and fishing journeys are usually short (usually less than one day and very often only a few hours). In general, in tropical fisheries the artisanal fleet land a variety of species, although there are examples of the use of selective fishing gear. In temperate and cold climates artisanal fleets can focus more easily on specific species according to the period of the year; nevertheless, they may land a variety of species to respond to the market demand.

Although very often artisanal fisheries are seen as an unsophisticated practice, closer scrutiny will reveal that in many cases they are passing through a process change. There are many reasons for this process but very often the main driving forces are: urbanization, fish exports and competition with the industrial fleet. This change in the scenario of artisanal fisheries is essential to understanding the fish handling problems faced by the artisanal and small sector of the fish industry, particularly in developing countries.

When the artisanal fleet was serving small villages, the amount of fish handled was very low; the customers usually bought the fish direct from the landing places, fishermen knew customers and their tastes, and fish was consumed within a few hours (e.g., fish caught at 06.00 h, landed and sold at 10.00 h, cooked and consumed by 13.00 h). In this situation, ice was not used, and gutting was unknown; very often fish arrived at landing places in rigor mortis (depending on fish species and fishing gear), and fish handling was at most reduced to covering the fish from the sun, keeping it moist and keeping off the flies.

With urbanization and the request for safer and more quality products (as a result of exports and competition with industrial fish) conditions changed drastically. Large cities also demanded increased fish supplies, and thus middlemen and fish processors had to go to more distant landing places for fish. The amount of fish handled increased, fishing journeys lasted longer and/or passive fishing gears like gillnets were set to fish for longer times, a chain of middlemen and/or official fish markets replaced the direct buyer at the beach and, as a result of growing business (fish for income), in some places the catch effort also increased with a consequent increase in the number of fishing boats and an increase in the efficiency of the fishing gears. In one way or another, each of the new circumstances added hours to the time which passed between catching the fish and eating or processing it (e.g., freezing). This increase in exposure of un-iced fish to ambient temperature (or water temperature for a dead fish in a gillnet), even though brief (e.g., an additional 6-12 hours), dramatically changed the situation regarding fish spoilage and safety.

In the new situation, fish remained at ambient temperature some 13-19 or more hours. It could be already spoiled, at terminal quality and/or could present public health hazards (e.g., from the development of C. *botulinum* toxin to histamine formation). In addition to the safety and quality aspects, post-harvest losses, non-existent at subsistence level and very low at the village stage, become important. For instance, it is estimated that the post-harvest losses of Nile perch caught artisanally in Uganda amount to 25-30% of the total catch.

The situation described in previous paragraphs, and cases like, moved extension services in developing countries and international technical assistance to focus on the problem of introducing improved fish handling methods at the artisanal level. The basic technical solution is the introduction of ice, proper fish handling methods and insulated containers, which is the approach utilized by most of the artisanal fleet in developed countries. There are several examples where this approach was adopted by fishermen in developing countries and has become a self-sustained technology. Two very interesting cases to analyze are the introduction of insulated containers onboard of "navas", the traditional fishing vessels of Kakinada in Andhra Pradesh, India (Clucas, 1991) and the introduction of insulated fish containers in the pirogue fleet of Senegal. The sketch of an insulated fish container for Senegalese pirogues.

The type of catch and needs expressed by fishermen. The materials and tools needed to construct the insulated container are available to fishermen in Senegal, even though some of them are imported (e.g., foam sheets and resin). The example of Senegalese fishermen is now spreading steadily to similar fisheries in Gambia, Guinea-Bissau and Guinea which are adopting the use of insulated containers similar to those of Senegal. However, the process of diffusion and adoption of a technology, even if relatively simple, is not as straightforward as could be supposed. A pirogue with two insulated containers onboard.

Once artisanal fishermen become aware of the rationale of insulated containers, they tend to favour large insulated fish containers rather than small ones. The reason is clear from Equations 7.e and 7.g, as for the same volume of fish and ice, large containers will present less external area than the area presented by several small containers. For example, a large cubic insulated fish container can be envisaged of a side measuring x m, and eight cubic insulated containers of sides equal to x/2 m presenting the same total volume as the large one. The eight containers will have an external area twice that of the big container, thus increasing the ice consumption by two, and decreasing the amount of fish that can be transported.

Other reasons are that small containers will cost more than a large one of the same total volume (simply because they need more material); small containers are not always easy to secure safely onboard small boats, and large containers allow for transport of large ice bars that can be crushed at sea (reducing stowage rate). However, large containers are difficult to handle and sometimes canoes and pirogues are very small or narrow and they cannot accommodate large insulated fish containers. This is the case for relatively small insulated fish containers. A serious constraint in many artisanal fisheries is the relatively high cost of industrial containers and the difficulty in finding appropriate industrial materials to construct them. For this reason, efforts have been made to develop artisanal containers made from locally available materials.

In some cases, the correct approach could be to add insulation to local fish containers; in other cases it could be necessary to develop a new container. In general, artisanal fish could be cheaper than industrial fish containers, but they will not last as long.

An artisanal insulated container developed at Mbegani (Tanzania), based on the local basket container ("tenga"). A key factor in the construction of artisanal insulated containers is the selection of insulation material. There are a number of materials available: inter alia, sawdust, coconut fibre, straw, rice husks, dried grass, old tires and rejected cotton.

However, the use of such materials presents problems: the materials become wet very quickly (with the exception of old tires), losing their insulating capacity and increasing the weight of the container. When wet, most of them tend to rot very quickly. The solution is to put them inside a plastic bag (waterproof); however, in this case they tend to settle, leaving part of the walls without insulation.

With a view to overcoming these problems, the concept of "insulated pillows" was developed in various FAO/DANIDA fish technology workshops. This concept is very simple: the insulating material (e.g., coconut fibres) is placed inside one plastic tube of the type usually found to produce ordinary small polyethylene bags (10 cm in diameter); the insulating material is pressed before sealing the tube; the tube is sealed by heat at both ends (e.g., every 20 cm), and with some practice it is possible to produce a strip of "pillows". It is advisable to utilize a second tube to reduce the incidence of punctures due to fish spines and bones.

The strip of "insulated pillows" can then be placed between the internal and the external walls of the container. Once the container is finished with an insulated lid and handles, fish and ice can be put in a large resistant plastic bag. The use of the plastic bag extends the lifespan of the container and improves fish quality. This example indicates the type of practical problems found when developing an artisanal insulated fish container, and the possible solutions.

Chill fish when necessary

Despite the knowledge on the advantages of fish chilling, ice it is not as widely used as it should be, particularly at artisanal level in developing countries. Which are the main problems found in practice? Some of the problems that can be found are as follows:

Ice should be produced mechanically

This obvious statement implies, *inter alia,* that it is not possible to produce ice artisanally for practical purposes (machines and energy are required). To produce ice under tropical conditions, from 55 to 85 kWh/l ton of ice (depending on the type of ice) are necessary whereas, in cold and temperate countries from 40 to 60 kWh are required for the same purpose. This may be a large power requirement for many locations in developing countries, particularly in islands and places relatively far from large cities or electricity networks. Ice plants require maintenance and hence trained people and spare parts (in many cases this requires access to hard currency).

A cold chain will also require chill rooms (onboard and on land), insulated containers, insulated trucks and other auxiliary equipment (e.g., water treatment units, electric generators). Besides increasing the cost, all this equipment will increase the technological difficulty associated with the fish cold chain.

Ice is produced and used within an economic context

In developed countries ice is very cheap and costs only a fraction of the price of fresh fish. In developing countries ice is very often expensive when compared with fresh fish prices.

A survey conducted in 1986 by the FAO/DANIDA Project on Training on Fish Technology and Quality Control on current fish and ice prices in fourteen African countries demonstrated that in all cases and for all the fish species, I kg of ice increased the fish price at least twice the rate recorded in developed countries. The cheaper the fish the worse the situation. For instance, in the case of small pelagics, the percentage of increase in the fish cost per kilogramme of ice added, was 40% for the "yaboy" of Senegal, 16-25% for the sardinella of Congo, and 66 % for the sardinella of Mauritania and the anchovy of Togo. The market price for fish, in this case, acts as a deterrent for the use of ice.

According to the relative cost of ice to fish, ice may or may not be used. For instance, in Accra, Ghana in 1992, it was found that using ice to chill small pelagics (Ghanian herring) in a proportion of 2 kg ice: 1 kg fish would increase the cost of fish by 32-40%. However, in the case of snapper, for the same ratio of ice to fish the cost increase would be in the range of 4.5-5.7%. The result is that ice chilling of snapper is relatively common in Accra, whereas ice is not utilized to chill small pelagics. Very often fish compete with other sources of demand (soft drinks, beer), even if the ice machine was initially installed to supply ice for chilling fish. This and energy losses at the ice plants contribute to increase the market price of ice. In addition to producing and utilizing ice on a sustainable basis, economic aspects must be considered (e.g., depreciation, reserves, investment). Moreover, in the case of ice manufacture there is a strong influence of the scale of production. Low ice prices in developed countries are also the result of large ice plants located at the fishing harbours that supply a large number of companies and fishing boats.

Practical constraints

Introduction of ice into fish handling systems that are not accustomed to using it can create practical problems. It is clear that the introduction of ice will increase the volume required for storage and distribution, and will reduce the effective fish hold in vessels. The use of ice will also increase the weight to be handled. This will have a number of implications such as an increased workload for the fishermen, fish processors and fishmongers, and an increase in costs and investment. It is clear that the total amount of ice needed per 1 kg of fish, in the complete cycle from the sea to the consumer will be much higher in tropical countries than in cold and temperate regions. As an indication, the average consumption of ice in the Cuban fishery industry was estimated at around 5 kg of ice per 1 kg of fish handled (including ice losses), although higher values (up to 8-10 kg of ice per 1 kg of fish) have been recorded in single industries in tropical countries; this necessitates large storage and transport capacities.

Freshwater or seawater utilized for producing ice should comply with standards (microbiological and chemical) for potable water and should be readily available in the volumes required. This is not always possible particularly in countries with energy problems (blackouts) and without (or with erratic) public tap-water distribution. If water has to be treated, this implies additional costs and additional equipment to operate and maintain. Properly trained personnel are required to operate the ice plant and auxiliary equipment efficiently, and to handle ice and fish properly. Although many developing countries have made efforts to train people, in many cases there is a lack of technical personnel ranging from well trained fish technologists to refrigeration mechanics or electricians, or simply plant foremen. Moreover, in many developing countries it is increasingly difficult to keep technical and professional schools operating in this field, thus jeopardizing the possibility of self-sustained training, and hence fishery industry developments.

Ice is not an additive

Knowledgeable people (e.g., fishmongers) are quickly aware of the fact that ice is not an additive. Therefore, when there is a delay in icing, ice is not usually utilized (even if available) because it will not improve fish quality.

Consumers could also be intuitively aware of this fact, and they prefer to be presented with the fish as it is (e.g., at the terminal state of its quality) rather than in ice, because in this case ice will increase the price of fish but not enhance its quality. Due to the above and to the problems associated with the transition between artisanal and industrial or semi-industrial fisheries, already discussed, consumers in some countries (e.g., in Saint Lucia and Libya) tend to believe that iced fish is not fresh fish.

A need for chilled fish can develop if a market for iced fish (not just a market for "fresh fish") is developed, and to develop a market for iced fish where it does not already exist may be a very difficult and expensive endeavour as is the introduction of any other food product.

Need for appropriate fish handling technologies

To chill and keep fish with ice is a very simple technique. A more complicated picture emerges when actual fish handling systems are analysed, including the economic aspect. From a comparative study on the same fish handling operation, utilizing ice and insulated containers, carried out in both a developed and a developing country, it was seen that in developed countries, the more "appropriate" technology would aim at reducing wage costs (e.g., chutes to handle ice and fish, special tables to handle containers and boxes and conveyors to move them, machines that mix ice and fish automatically); in developing countries the main concern would be to reduce ice consumption, and to increase the fish : ice ratio in the containers (Lupin, 1986 b).

The same study found that a twentyfold difference in wage costs between developing countries and developed countries cannot offset a tenfold difference in the cost of ice. There is no "comparative advantage" in low wages in developing countries with regard to fresh fish handling. Advanced technology on fish handling from developed countries could make work easier for people in developing countries, but might not improve the economics of the operation as a whole.

There is obviously no single solution to the problems discussed above. However, it is clear that it is the problem to be solved in the coming decade in the field of fresh fish handling. With total catches having reached a plateau, losses due to the lack of ice utilization could be ill-afforded, and developing countries and artisanal fishermen in particular should not be deprived of potential market opportunities.

IMPROVED CATCH HANDLING IN INDUSTRIAL FISHERIES

The aims of modern catch handling are the following:

- to maximize the quality of the landed fish raw material. It is of particular importance to provide a continuous flow in handling and to avoid any accumulation of unchilled fish, thereby bringing the important time-temperature phase under complete control.

- to improve working conditions onboard fishing vessels by eliminating those catch handling procedures which cause physical strain and fatigue to such a degree that no fishermen need to leave their occupation prematurely for health reasons.
- to give the fisherman the opportunity to concentrate almost exclusively on the quality aspects of fish handling.

To meet these aims, equipment and handling procedures that will eliminate heavy lifting, unsuitable working positions and rough handling of fish must be introduced. By doing so, the catch handling time is accelerated and the chilling process initiated much earlier than was previously the case (Olsen, 1992). Important general aspects in modern catch handling are:

- phase one, which covers the time used for the necessary handling onboard, i.e., the time until the fish is placed in chilling medium, must be as short as possible. The fish temperature at time of capture can be high with consequent high spoilage rate.
- phase two-the chilling process-must be arranged so that a fast chilling rate is obtained for the whole catch. Maximum chilling rate will be obtained by a homogeneous mixing of fish and ice, where the individual fish is completely surrounded by ice and the heat transfer therefore is maximum, controlled by the conduction of heat through the meat to the surface. This ideal situation can be obtained during chilling of small pelagics in a chilled seawater (CSW) system; but by chilling demersal food fish in boxes with ice it is not always possible to obtain homogeneous fish/ice mixing. However, the appearance of fish completely surrounded by ice is often deteriorated due to discolorations and impression-marks. In practical life, icing is therefore often done by placing a single layer of fish on top of a layer of ice in the box even if it is bad practice from a temperature control and therefore shelf life point of view. Cooling is primarily achieved by melt-water dripping from the box stacked on top. This type of chilling will only function satisfactorily if fish boxes are shallow and have a perforated bottom.
- in phase three, which covers the chilled storage period, it is important that a homogeneous temperature at-1.5°-0°C is maintained in the fish until first hand sale. As this period may be extended for several days, this aspect has top priority.

Catch handling can be done in several ways ranging from manual methods to fully automated operations. How many operations will be used in practice and the order in which they are done depends on the fish species, the fishing gear used, vessel size, duration of the voyage and the market which has to be supplied.

Transferring catch from gear to vessel

Midwater trawlers and purse seiners fishing pelagic fish use tackling in lifts of up to 4 t, pumping or brailing for bringing the catch onboard. When lifting huge hauls (100 t or more) onboard by these methods, the danger of losing fish and gear always exists if the fish start to sink after having been brought to the surface. The speed of which the fish may sink depends on the species, catching depth and weather conditions during hauling. Pumping the catch onboard using submersible pumps without bruising the fish can be difficult, as it is not easy to control the fish-to-water ratio during pumping.

In recent years, the so-called P/V pump (P/V-pressure/vacuum) has found increasing use. The P/V-pump principle is that an accumulation tank of 500-1500 1 size is alternately put under vacuum and pressure by a water-ring vacuum-pump. The fish, together with some water, are sucked through a hose and a valve into the tank of the system. When the tank is full, it is pressurized by changing the vacuum and pressure side connections from the tank to the pump and the fish/water mix flows through a valve and a hose into a strainer. The P/V-pump is claimed to handle the fish more gently than other fish pump types, but the capacity is generally lower, mostly because of the alternating way of operations. This problem can be solved by having two P/V-tanks running in phase opposition using only one vacuum-pump.

Small gillnetters (10-15 m) haul the nets with the net hauler, and very often store their catch in the net until landing. Here the net is drawn through a net shaker by two men in order to free the fish from the gear. It has been shown that the violent way in which the shaker works can be harmful to the men's hands, arms and shoulders. Ergonomic precautions have therefore been suggested to overcome this problem.

Trawlers and seiners (Danish and Scottish) tackle the catch into pounds. Commonly used pounds are those with a raised bottom which can be hoisted hydraulically. The purpose of these designs is to provide good working conditions for the crew. Also gillnetters may use a work-saving pound system, which is often connected with a conveyor to bring fish to the gutting-table.

Holding of catch before handling

When large catches are to be handled, or if for other reasons catch handling cannot start immediately, it is convenient and necessary to prechill the catch during holding in deck-pounds using ice or in tanks using Refrigerated Sea Water (RSW) or a mixture of ice and sea water (Chilled Sea Water, CSW). Prechilling holding systems are mostly used on pelagic trawlers which grade their catches in size before storing in boxes or in portable CSW-containers. It is also essential to prechill when the pelagic fish are soft and feeding and therefore very prone to bellyburst. Prechilling tanks are unloaded by elevator or P/V-pumps. If no sorting is done onboard, the fish is conveyed directly for chilled storage in the hold. A system for holding demersal fish in tanks.

Sorting/grading

Pelagic fish are sometimes sorted or graded onboard according to size. The equipment used operates on the basis of thickness of fish using principles such as:

- vibrating, inclined diverging bars
- contrarotating, inclined, diverging rollers
- diverging conveyors where fish are being transported along a power driven V-belt.

Grading by thickness can meet the demand for the high capacity needed in pelagic fish handling, but it is generally accepted that the correlations between thickness and length or weight are not too good (Hewitt, 1980). The most important point, often forgotten, for making a grader function at its optimum is even feeding. This could be done with an elevator delivering to a (vibrating) water sprayed chute leading to the inlet guide chute of the grading machine. Sometimes it is necessary to install a manual sorting conveyor before the grading machine for removal of larger fish and debris, e.g., in the fishery for argentine with by catch of grenadier.

Sorting and grading of demersal fish by species and by size is normally done by hand. However, some automatic systems of sorting according to width are in use. Static or dynamic weighing by marine weighing systems are also in use with good results. Research is under way using a computerized vision system for species and size grading.

Bleeding/gutting/washing

In order to obtain optimal quality in a white fillet, many white-fleshed demersal fish (but not all) need to be bled and gutted immediately after capture. The best procedures from an economic, biological and practical point of view are still under discussion. The vast majority of fishermen are handling the fish in the easiest and also the fastest way, which means the fish are bled and gutted in one single operation. This may be done manually, but gutting machines have been introduced to obtain even more speed. The fish are transported to and from the fisherman by suitable conveyor systems. Using machines, round fish can be gutted with a speed of approximately 55 fish/minute for fish length up to 52 cm and 35 fish/minute for fish length up to 75 cm. Gutting by machine is 6-7 times faster than hand-gutting.

Existing gutting machines for round fish of the type using a circular saw blade for cutting and removing the guts destroy the valuable roe and liver. A new type of gutting machine which copies the manual gutting procedure is now available on the market. Gutting speed of this machine is 35-40 fish/minute and the roe and liver can be saved (Olsen, 1991). Flatfish can also be gutted by a recently developed machine. The speed of this machine is about 30 fish/ minute. After gutting, the fish are conveyed to the washing or bleeding operation. This may be done in pounds, often with raised bottom or in special bleeding tanks, frequently with a hydraulically-operated tilting system and rotating washing drums are also used; and special equipment such as the Norwegian and British fish washer may be used.

After catch handling (sorting, grading, gutting, etc.), the fish may be passed to an intermediate storage silo or batch holding system for the different sizes or grades before being dropped by chute to the hold, or the chutes may lead directly from the grading machines to the hold.

Chilling/chilled storage

Demersal fish have traditionally been stored on shelves or in boxes. Boxing has a big advantage over shelf storage as it reduces the static pressure on the fish and also facilitates unloading. Shelf storage is done by alternating layers of ice and fish from one layer of ice and fish (single shelving 25 cm between shelves) up to ice/fish layers 100 cm deep. In practice, shelving often allows better temperature control than boxing and therefore also a longer storage life. Because excessive handling during unloading and excessive pressure on the fish have-a negative effect on quality, e.g., appearance, boxing is preferable to shelving, given proper icing.

In pelagic fisheries, boxed fish will be untouched until processed, but in demersal fisheries the catch is often only sorted by species onboard and not graded by size and weighed. These operations are carried out after landing before auction whereby some of the handling and quality advantages of boxing are lost.

In the near future when integrated quality assurance systems have been introduced, these unit operations will be carried out onboard and an informative label on each box will give details of factors of importance for first-hand sale (including freshness). In general, two types of plastic fish boxes are used: stack-only and nest/stack boxes.

To overcome some of the space problems in using stack-only boxes, the nest/stack type has been developed. These occupy only approximately a third of the space needed when stored empty compared to when the boxes are loaded with fish and ice.This type of box is widely used in France, the Netherlands and Germany and also in some Danish ports. When a system tailor-made for a certain type of plastic box is designed, the quality advantages of using boxes can be fully utilized onboard. The key points to consider are:

1. The handling rate necessary to prevent quality loss because of delayed icing. Prechilling can be of advantage to compensate lack in handling rate.
2. Handling methods which make it possible to guarantee that the icing procedure is sufficient to chill the fish to 0°C and maintain this temperature until landing.
3. The hold construction must be constructed such that safe and easy stacking of the boxes can take place.
4. Hold insulation of a relatively high quality should be considered. A small mechanical refrigeration plant can be of advantage. Air temperature in the hold should be + 1°-3°C

RSW-storage (Refrigerated Sea Water is a well established practice which has been refined both theoretically and practically since its introduction in the 1960s in Canada where it was developed for salmon and herring storage (Roach et al, 1967). At the beginning, most RSW vessels were salmon-packers and because of some failures in design which were attributed either to insufficient refrigeration or circulation systems, a standard for control of RSW-systems was established. Since vessels are different, the RSW-installation has to be studied carefully in every fishery to determine its real capability. Therefore, methods for rating each individual system and vessel and providing general specifications and guidelines for the proper installation have been suggested by the Canadian technicians.

In order to obtain maximum shelf life from RSW-systems, temperature homogeneity in the region of-1°C is very important. The factors affecting temperature homogeneity were recently studied in Denmark (Kraus, 1992). The most important conclusions were that the inflow of the chilled seawater in the bottom of the tank must take place over the whole tank bottom area, and that filling capacity for securing water circulation and temperature homogeneity is dependent on fish species. The necessary chilling rate was suggested to be: fish temperature must be below 3°C within four hours and below 0°C after 16 hours, and the temperature should be kept between-1.5°C and 0°C until unloading. The CSW system has also been developed in Canada as a much cheaper means an investment point of view-to obtain rapid uniform chilling of fish. The most popular method used is the so-called "Champagne" method where rapid heat transfer between fish and ice is obtained by agitation with compressed air introduced at the bottom of the tanks, instead of using circulation pumps as in RSW and some earlier CSW designs. An indication of the chilling rate for herring could be: reduction of fish temperature from 15°C to 0°C within two hours. The concept of a CSW system is to load well insulated tanks at the harbour with the amount of ice necessary to chill the catch to between 0' and-1°C and maintain this temperature until unloading.

The Canadian west-coast fishermen are achieving this in practice by using a minimum of seawater when they start loading the tank and by forcing air through the ice-sea-water-fish-mixture only during loading, and stop forcing air immediately when the tankis full. Thereafter they will force the air only for 5-10 minutes with 3-4 hours' interval. The air agitation therefore only serves as a method to overcome local temperature differences in the tank. The objective is to obtain a uniform mixture of fish and ice in order to secure temperature homogeneity.

A proven rule-of-thumb for estimating the amount of ice necessary is simply to observe the amount of ice left in the tank at unloading, and compare it with temperature readings, which should be in the-1°C range measured in the landed fish. The starting situation should be conservative, which at sea-temperature around 12-14°C, for a trip lasting 7 days and with 10 cm polyurethane insulation, is 25% ice by weight of the tank capacity. The amount of ice is adjusted according to the observations on the following trips.

An analytical approach to estimate necessary ice quantities in a CSW tank system has been developed. The quantity of ice required takes into consideration tank size, catch volume, time at sea, water temperature, hold insulation and hold flooding strategy.

CSW "Champagne" systems can also be used in small coastal vessels, e.g., in a fishery for small pelagic fish with vessels of 10-14 m length with a fish carrying capacity from 3 to 10 t fish.

Another way of loading a CSW tank, which is in practical use in Denmark, is to add the necessary amount of ice to the fish during loading by mixing a controlled stream of fish with a controlled stream of ice. The greatest amount of ice is added to the fish during loading. When the tank is full the voids are filled with seawater from a hose and the tank is left undisturbed, except for watercirculation by pumping or compressed air blowing for 5-10 minutes of 4-hour intervals. The ice is bulk-stored in the forward hold and the ice is shovelled into a conveyor flush with the floor. The conveyor then leads the ice to the mixing point at the deck.

The use of portable CSW containers for pelagic fish handling was tested in the early 1970s. The approximately 2 m^3 heat insulated containers were loaded with the necessary amount of ice from the harbour and agitated with compressed air in a similar way as for CSW-tanks. The main advantages with this method are that the fish will be undisturbed until processed and easily unloaded. The disadvantages are: marketing problems and reduced pay-load on existing vessels (Eddie, 1980). Portable 1.1m^3 CSW containers are used to a limited extent in combination with the earlier mentioned conveyor system originally laid-out for boxing without the above-mentioned reduced pay-load compared to boxing (Anon., 1986). Also, small coastal vessels can use insulated portable CSW containers.

Unloading

Shelfed fish are unloaded, using baskets or boxes which are filled as the shelves are removed. The fish are tackled from the hold and emptied on a conveyor leading to the manual grading and weighing process. Boxed fish iced in 20 or 40 kg boxes at sea will normally be unloaded in pallet loads of, for instance, twelve 40 kg boxes per pallet. Swedish boats use hydraulic deck-mounted cranes and a special pallet fork during unloading. An unloading rate of approximately 30 t/h is possible by this method.

Danish coastal vessels, landing their pelagic catches daily, use quay mounted P/V-pumps for unloading their catches, which often are iced in pens in layers up to approximatly 1 m height. It is necessary only to add small quantities of water to make the pump function properly. The fish is delivered to a strainer from where a conveyer leads the fish to a size grader. The strained water is recirculated to the hold. Grading machines with up to 30 t/h are often installed. In Scandinavia the 30-50 in RSW/CSW vessels still use brailing to a limited extent when unloading their catches at a rate of 30 to 50 t/h. The main disadvantage of this method is that very big hatches are needed to obtain reasonable unloading rates.

P/V-pumps have recently been introduced for unloading herring and mackerel. Thus vessels with small tanks, e.g., 30 in, and small hatches can also be unloaded at a rate similar to or higher than the above-mentioned brailing rate. P/V-pumping rates will typically be around 40-50 t/h. The fish can be transported directly in a tube system into the factory where representative samples are taken for quality assessment.

FISH AND GENETICS ADVANCE BLADDER-CANCER DIAGNOSIS

In the United States, bladder cancer is the fourth most diagnosed malignancy for men, and the tenth most common for women. Despite the fact that diagnosis rates have leveled off over the past decade, bladder cancer is still of great concern to the 60,000 people who are estimated to be diagnosed with the disease this year. Yet, the disease rarely receives the high-profile attention of other statistically similar cancers, such as melanoma.

There are three main types of bladder tumors, identified by the way they appear under the microscope:

- urothelial carcinoma (also known as transitional cell carcinoma or TCC);
- squamous cell carcinoma; and adenocarcinoma.

These same types of cancer, can also grow in the ureter and the urethra, which is why a complete evaluation of the urinary system is recommended for patients diagnosed with a cancer of the bladder, ureter, or urethra.

Cancer in the urothelial cells lining the bladder (also known as urothelial or transitional cell carcinoma) is the most common form of bladder cancer, accounting for 90% of diagnosed cases. Only about 2% of bladder cancers are squamous cell carcinomas, which look much like skin cancer under the microscope, and only about 1% to 2% of bladder cancers are adenocarcinomas. While nearly all squamous cell cancers and adenocarcinomas of the bladder are invasive, urothelial tumors can be categorized as noninvasive or invasive. In noninvasive cases, the cancer has not spread further than the urothelium, which is the innermost layer of the bladder. In invasive cases, however, the malignancy spreads through the lining of the bladder into the deeper tissues of its wall, and even into the thick, deep bladder muscle called the muscularis propria. These invasive cases are often far more serious, and outcomes are much less certain.

3

Sustainable Development of Fisheries

BACKGROUND

In fisheries, development may be defined as "a process of change through which sustainable and equitable improvements are made to the quality of life for all or most members of a society" (Bailey and Jentoft, 1990). Adding the need for sustainability implies that these improvements need to be achieved without risk to the long-term stability of the ecosystem concerned.

In 1987, the World Commission on Environment and Development defined sustainable development simply as "Development that meets the needs of the present without compromising the ability of future generations to meet their own needs." This definition recognises that human needs can change with time, and that preservation of the integrity of the environment is necessary to ensure that future generations can realise the full potential of their own development needs. Another, more elaborate definition states that "Sustainable development is the management and conservation of the natural resource base, and the orientation of technological and institutional change in such a manner to ensure the attainment and continued satisfaction of human needs for present and future generations.

Such development conserves land, water, plant and genetic resources, is environmentally non-degrading, technologically appropriate, economically viable and socially acceptable." This definition was developed by the FAO Committee on Fisheries in 1991, and acknowledges the need to achieve a workable balance between guaranteeing satisfaction of both present and future human needs including fulfilment of social and economic demands, and conserving the natural resource base. In response to the general goals identified, FAO developed a Code of Conduct for Responsible Fisheries (FAO, 1995) that provides principles for ensuring sustainable exploitation of marine resources.

SEARCHING FOR OPTIMUM FISHERIES MANAGEMENT

The 1991 FAO definition noted earlier recognises the need for management of the resources concerned as one of the essential actions for achieving sustainable development. To accommodate the other components of the FAO definition, it is clear that such management must embrace a range of objectives, incorporating biological, economic, and social elements. Knowledge of the status of the resources and the likely impact of fishing activities and other factors on these resources provide for informed and successful management of fisheries.

In the early days of fisheries management, scientists focused on the biological modelling of fish stocks (Graham, 1935; Beverton and Holt, 1957). Assuming that fish stocks respond predictably to moderate levels of exploitation, with a defined equilibrium state, fisheries scientists are able to calculate the maximum level of catch that would allow sustainable exploitation. This sustainable maximum catch is referred to as the Maximum Sustainable Yield (MSY), and continues to be a key reference point in present-day fisheries management (Gulland, 1983).

MSY takes into account only the sustainability of the resource and so satisfies only a biological objective. In view of this, fisheries economists have argued that fishing is conducted as a business and that the economic benefits of the activity must not be ignored. This led to the development of another management reference point, referred to as Maximum Economic Yield (MEY). MEY is calculated based on optimising the difference between the cost of fishing and the income gained, and usually occurs at a lower fishing effort than the fishing effort occurring at MSY (Hersoug, 1996).

Subsequent to the development of MEY, social scientists have noted the importance of addressing social concerns, to ensure more equitable distribution of economic benefits. The social objective of Maximum Social Yield (MsocY) was then recognised, and the associated reference point varies with the fishery situation.

These biological, economic and social management objectives, as currently defined, are now facing yet another challenge, with the development of chaos theory (Smith, 1990 in Symes, 1996; Wilson and Kleban, 1992 in Symes, 1996). Chaos theory suggests that although nature is non-random, it is unpredictable. The equilibrium state of fish populations assumed by MSY is not accepted, and fish populations are believed to vary unpredictably within limits (Symes, 1996). Chaos theory therefore proposes 'ecologically adapted management', based on knowledge of the long-term stable ecological relationships, and requires more flexible management systems.

ACHIEVEMENT OF RESPONSIBLE FISHING

Sustainable fisheries development can be achieved through responsible fishing, which considers rational fishery management objectives that address a range of issues including the status of the resource, the health of the environment, post-harvest technology and trade, as well as other economic concerns, social benefits, legal and administrative support. In the case of shared resources, a co-ordinated approach to responsible fisheries management is essential, and Caddy and Griffiths (1995) proposed the following actions:

Regulate Fishing Effort

It is crucial to control fishing effort, and to avoid financial incentives that would contribute to excess fishing capacity. Excess fishing capacity and overcapitalization threaten the sustainability of the resource, as well as the industry.

Establish Code of Conduct for Responsible Fishing to Guide Management Plan

This is needed to maximize benefits from the fishery, while avoiding wastage caused by indiscriminate fishing practices. In particular, it is important to reduce catches of undersized fish and non-target species, and to avoid use of gears that have a negative impact on the environment.

Establish and support regional/international fishery commissions and organisations concerned with management of shared resources

The successful management of shared resources requires effective coordination by all the countries concerned. The relevant international fisheries agreements promote participation in, and financial support for, the work of these commissions and organisations.

Regular Consultation among Harvesting Countries

Parties sharing the resources need to consult and collaborate regularly so as to promote understanding and full cooperation. Set agreed management objectives and related reference points, incorporating a precautionary approach

Agreement on management reference points during the early stages of the fishery will help to ensure full cooperation of participants with management decisions. Where there is scientific uncertainty, a precautionary approach to management is recommended.

Develop Contingency Plans

Management plans should incorporate some contingency for dealing with sudden and unpredictable environmental changes caused by man-made or natural disasters.

Develop Mechanism for Resolving user Conflicts

Management should provide mechanisms for handling problems arising from resource user conflicts.

Protect Biodiversity

The biodiversity or species richness of an ecosystem is an important measure of ecosystem health. The preservation of biodiversity will ensure that present human development activities do not threaten the ability of future generations to meet their own needs.

Protect the Environment

There should be monitoring and control of waste disposal and pollution. In addition, every effort should be made to prevent discarding of entangling material that could trap and kill species or physically damage the environment.

Promotion of Research

Research should be conducted to support and inform various aspects of management.

Optimise Social and Economic Stability

There should be fair and equitable distribution of benefits derived from the fishery.

Constraints to Caribbean Island States

For many small developing island states within the Caribbean, there are a number of constraints that pose significant challenges to the sustainable development of fisheries, including the following:

- *Fishing Practices:* Destructive fishing methods such as dynamiting fish are still practiced in certain areas. Additionally, fishers continue to use illegal mesh sizes that catch very small fish, and non-target species that may have little or zero market value.
- Inadequate knowledge of the resource and ecosystem
  Many Caribbean islands do not have sufficient detailed data to permit an accurate evaluation of the status of their resources and a good understanding of their marine ecosystem.
- Inadequate knowledge and recognition of social and economic conditions
  Many governments within the region still do not have an adequate appreciation of the social and economic potential of sustainable fisheries development, and hence still invest minimum resources in fisheries development and management activities.

- *Absence of Long-term Policies:* The lack of long-term policies makes it difficult for Caribbean countries to maintain pace with rapidly evolving trends in global fisheries management approaches and trade strategies.
- *Capacity:* In many Caribbean countries, fisheries administrations are poorly staffed, and there are limited numbers of skilled and knowledgeable fisheries technicians, scientists and managers, and limited equipment and funds available for basic tasks such as data collection and research. There are few research institutions and regional organisations within the region, and some of these suffer similar limitations in available funding and a broad range of technical expertise.
- *Resource user Input:* In the Caribbean, fisherfolk organisations have not actively participated in the management process, mainly owing to a lack of good organisational and administrative skills. On the part of the fishers, this has resulted in a lack of trust in, and respect for, the governments that are responsible for making fisheries management decisions.
- Capability for monitoring, control and surveillance
  In many instances, there is little or no capacity for monitoring, control and surveillance activities, and no associated legislative framework.
- *Post Harvest Aspects:* Caribbean countries are just beginning to develop their export markets for fish and fish products, and face the challenge of satisfying stringent standards in all aspects of fish handling, processing and packaging, recently established by the importing countries.
- *Facilities:* Equipment and facilities are not upgraded regularly to deal with expanding fishing activities and increasing management demands.
- *Habitat Degradation and Pollution:* For some time, there have been signs that the Caribbean ecosystem is under stress (Richards, and Bohnsack, 1990). Degradation of habitat, particularly in the coastal areas has resulted from uncontrolled coastal development activities mostly associated with expanding tourist industries and overpopulation problems, as well as extensive sand mining and deforestation.
- *Financial Resources:* Many Caribbean countries are small island developing States with limited financial resources. This impacts negatively on their ability to cope with the wide range of issues required for successful sustainable fisheries development.

Caribbean governments need to appreciate the social, economic and financial potential of sustainable fisheries development. Fisheries administrations need to be given more financial resources, and properly staffed and equipped to address effectively and completely all aspects of sustainable fisheries development and management. Additionally, small island states should recognize the benefit of, and work towards, regional coordination in fisheries management activities, including sharing of expertise and resources for education, research, technology, monitoring, control and surveillance activities, and development of the relevant legislative framework.

Having removed constraints to responsible fishing, fisheries management must then strive to achieve a workable balance between defined objectives protecting the resource and its environment, and those seeking fair and equitable distribution of viable economic benefits.

CURRENT PROBLEMS IN THE MANAGEMENT OF MARINE FISHERIES

The public perception of fisheries is that they are in crisis and have been for some time. Numerous scientific and popular articles have pointed to the failures of fisheries management that have caused this crisis. These are widely accepted to be overcapacity in fishing fleets, a failure to take the ecosystem effects of fishing into account, and a failure to enforce unpalatable but necessary reductions in fishing effort on fishing fleets and communities. However, the claims of some analysts that there is an inevitable decline in the status of fisheries is, we believe, incorrect. There have been successes in fisheries management, and we argue that the tools for appropriate management exist. Unfortunately, they have not been implemented widely. Our analysis suggests that management authorities need to develop legally enforceable and tested harvest strategies, coupled with appropriate rights-based incentives to the fishing community, for the future of fisheries to be better than their past.

The United Nations Food and Agriculture Organization (FAO), which monitors the state of world fisheries, has estimated that since 1990 approximately one-quarter of fish stocks have been overexploited, depleted, or are recovering from depletion (17%, 7%, and 1%, respectively), with the Northeast and Northwest Atlantic, the Mediterranean, and the Black Sea being the areas with the largest number of depleted stocks. Many authors have elaborated on these conclusions, documenting the poor state of fisheries worldwide. Nevertheless, the situation, although serious, is not catastrophic, and there are grounds for optimism. There have been successes of fisheries management, and there is an understanding of what is involved in successful fisheries management and of the requirements for its implementation. These issues are explored in this review.

The management of commercial fisheries clearly requires a good scientific understanding of the behavior of the exploited stock or stocks. The science that is used to assess commercially exploited species is still dominated by the population models developed by Beverton and Holt for single-species assessments some 50 years ago. The availability of substantial computing power has meant that sophisticated estimation methods can be used, and an appreciation of the way in which fish stocks respond to environmental variability is readily incorporated in scientific advice. Calls for a more ecosystem-orientated approach have been voiced for some while, but the paucity of data and the demands of multiparameterized multispecies models means that most ecosystem considerations in practical stock assessment tend to be ad hoc manipulations of the single-species approach.

What has developed is a realization that effective management requires an understanding of how the fishery system is performing relative to reference points. The most commonly used reference points are those relating to the size of the stock itself and the fishing mortality that will result in these stock sizes, given existing relationships between the stock, recruitment, natural mortality, and growth. A typical "target reference point" is the biomass necessary to produce maximum sustainable yield (BMSY). However, such targets do not explicitly recognize threats to the stock. To address this issue, stock size "limit reference points" are usually defined or interpreted as the stock biomass below which recruitment becomes substantially reduced. Clearly, it is important to avoid situations where the stock is at or below this level. Accordingly, management should aim to have as a target a level of stock size that carries a low risk (allowing for scientific uncertainty) of the stock dropping below the limit reference point. This could mean having a target level of fishing mortality that provides stock sizes above BMSY.

UNDERSTANDING FISHERIES MANAGEMENT

Competent scientific advice based on appropriate data is far from ubiquitous in the fisheries world, and even in ideal situations, fisheries management has often been unsuccessful. The success of a management system is often defined in terms of biological, economic, social, and political objectives. Clearly, economic and social objectives will not be met while a stock is in such a depleted state that the long-term sustainability of the fishery is threatened, but equally, biological objectives are unlikely to be met without consideration being given to economic and social objectives.

Hence, we argue that an understanding of the fishery management process can only come from analyzing the capacity and incentives of the two key stakeholders: the fishing community and the management authority. This is not to belittle the importance of other stakeholders, such as recreational fishers and environmental groups, who have important roles in the management of certain fisheries.

Where management is weak or nonexistent, the economic factors underlying overfishing in commercial fisheries have been generally understood since the 1950s. In short, when multiple fishers compete to catch fish from a given population, each fisher maximizes his net income by continuing to fish as long as the value of his catch exceeds the cost of catching it. An equilibrium, called the bionomic equilibrium, is reached only when fishing has reduced the fish population to a level at which catch rates are barely sufficient to cover the costs of fishing. The population is then maintained at this level through biological processes of natural growth and reproduction. Thus, if the price:cost ratio is high, the bionomic equilibrium will result in a low stock of fish, and hence a low annual catch level; two characteristic features of overfishing. In addition, the so-called economic rents (total revenue minus total costs) from the fishery will equilibrate at zero, resulting in minimal overall economic efficiency.

Many management authorities seek to meet their objectives by setting output controls in terms of a total allowable catch (TAC) for the year and closing the fishery when the year's cumulative catch has reached the TAC. Restrictions on fishing gear, fishing season, and fishing areas, as a supplement to the TAC, may also be imposed. If the TAC is correctly specified and enforced, this method should maintain a stock level well above that of bionomic equilibrium. If, however, the TAC and the science behind it are not respected by fishermen and not adequately enforced by authorities, widespread illegal fishing can occur. A recent example is in the eastern Baltic cod fishery where illegal fishing contributes to true catches being some 35 to 40% higher than reported.

Efficient enforcement can be difficult. Put simply, fishers will be deterred from breaking fishing regulations if their expected loss from detection and successful prosecution exceed their expected gain. In many fisheries, the probability of detection of illegal activity and the penalties are not sufficiently high to act as a disincentive.

Strong management can ensure that biological targets are met, but it is essential that regulations are enforceable, and this has often proved to be difficult. Less-than-perfect enforcement can lead to illegal fishing, poor scientific data, and a failure to meet biological targets. Input measures, such as limiting the number of vessels or restricting available season length, are usually more easily enforceable than output measures such as TAC. However, control via input measures is vulnerable to effort creep, whereby operators increase the fishing power of their vessels through technical means. Nevertheless, monitoring of vessel performance over time and adjusting the allowable level of effort have allowed successful effort control to be implemented.

Overcapacity

Simplistically, it would seem that positive economic rents should also emerge in a TAC-regulated fishery. In reality, many TAC-regulated fisheries have experienced an unexpected increase in fishing capacity, as additional vessels enter the fishery in response to (temporarily) positive rents. Economic models then predict a regulated bionomic equilibrium, in which economic rents (net of fixed costs) again equilibrate at, or near, zero. This situation currently exists in many of the world's regulated fisheries; overcapacity of fishing fleets is widely perceived as a major impediment to achieving economically productive fisheries. It is thus ironic that such overcapacity is usually generated by the management system itself, although it can also result from high profitability during the initial phase of a newly developing fishery.

Overcapacity is widely recognized as a major problem affecting world fisheries. With its attendant social and economic problems, overcapacity can, via the political process, lead to the erosion of management control. It is also understood to be one of the results of subsidizing fisheries, which even today is estimated to be several tens of billion U.S. dollars per year. Such subsidies directly undermine the sustainability of fisheries because they lead to a bioeconomic equilibrium with high levels of fishing and low stock size. In several fisheries, government funds have been used to buy out excess fishing capacity. For various reasons, such buyback programs have been less effective than expected. First, often only the least efficient vessels are bought up, leaving total fishing capacity largely intact. Second, the buyback program by itself does not remove the economic incentives underlying overcapacity, which tends to increase once the buybacks are completed.

Thus, the underlying cause of the dual crisis of overfishing and overcapacity, as well as other undesirable outcomes, such as habitat destruction and incidental kills of untargeted species, can be found in the economic incentives of fishers who compete for their annual catches. These incentives are not affected by management strategies that retain the competition between fishers for a common-pool resource. Perhaps the most important development in fisheries management over the past 20 years has been the recognition of this fact and the introduction of rights-based management in several regimes. Indeed, it has been argued that of the tools at the disposal of managers, more emphasis needs to be placed on incentive-based approaches that better specify community and individual harvest or territorial rights, in additional to public research, monitoring, and effective administrative oversight.

Transferable Quotas

An alternative management strategy based on individually allocated transferable annual catch quotas (ITQs, or individual transferable quotas) is now in effect in several fishing nations, including Australia, New Zealand, Iceland, Canada, and Namibia. A well-organized rights system alters the economic incentives of fishers, who no longer compete for their catches, so that highly competitive fishing no longer takes place. The guarantee to fishers of a certain proportion of the catch allows them to make rational economic choices about where and when they catch fish. An ITQ system goes further, allowing the industry to settle on a fleet capacity that optimizes individual economic yield to vessels or cooperatives, although this of course can still be distorted by inappropriate subsidies.

In addition, ITQ fishers may often be expected to favor management actions that protect and enhance fish populations, because the value of a quota share increases as stocks become more abundant. Problems that may arise, such as misreporting or high-grading of catches, have been successfully countered by the use of observers, required by the management system but paid for by the industry; observers are used extensively in the U.S. Pacific fisheries, Australia, and New Zealand. Experience with ITQ systems shows that many fishers willingly support and adhere to conservative management strategies and may also avoid fishing practices that endanger habitat or threaten other species, so long as they are guaranteed long-term rights. But this does not mean that enforcement and scientific monitoring are unnecessary in ITQ systems; both are essential unless catch levels are set at precautionary low levels. It is thus unsurprising that the two countries with perhaps the most fully developed ITQ systems, New Zealand and Iceland, have some of the highest costs of management per fishing vessel.

Several authors have pointed to instances of successful fisheries management in both the developed and developing world. Among their conclusions are that incentive structure, institutional capacity, and participation of stakeholders are of key importance. However, in some studies, a rights-based approach is seen as the primary mechanism to deliver this, whereas in others, severe top-down controls with very limited participation of fishing communities in the management process are advocated. We argue here that a necessary condition for successful management contains all these elements: a competent management authority able to set and enforce regulations and monitor the status of the stock, together with some form of rights-based allocation to fishing operators (either collectively or individually) to avoid the situation where overcapacity produces economic hardship and erodes management capacity.

EVIDENCE FROM FISHERIES PERFORMANCE

Reviews of successful fishery management are of necessity specific to individual fisheries and sometimes anecdotal. However, in some large areas, a combination of strong state governance and wealth, substantial scientific activity, and different types of fishery management offer the opportunity of some comparison between different types of fishery management that goes beyond the anecdotal.

Detailed data on the status of different fisheries are published for U.S., Northeast Atlantic, Australian, and New Zealand fisheries. The approach to management taken by these authorities is varied. New Zealand has the most developed and widespread application of individual user rights (ITQs), which have been in place from 1986 and have spawned other developments such as collaborative and alternative research by stakeholders. ITQs are present also in some Australian fisheries, a very few U.S. fisheries, and some Northeast Atlantic fisheries. (The Faroes, Norway, Iceland, and United Kingdom have rights-based systems, and some fleets, notably the Dutch flatfish and Spanish Grand Sole fleets, are also managed via ITQ.)

If a broad view is taken of these management areas, the evidence for the positive benefits of ITQs in supporting sustainable resource use is mixed. Only 15% of New Zealand's stocks within the quota management system, for which the stock status is known, are substantially below the target reference level.

For other administrations, the percentage of stocks that are below the limit reference level (i.e., overfished), out of the total number of stocks for which the status is currently known, is 19% for Northeast Atlantic fisheries managed by non-European Union (EU) administrations (Iceland, Faroes, and Norway), 25% for federally managed U.S. fisheries, 30% for Northeast Atlantic fisheries managed primarily by the EU, and 40% for Australian Commonwealth fisheries.

Even within the United States, there are very large regional differences: 40% of major Fish Stock Sustainability Index (FSSI) stocks managed by the New England and Mid-Atlantic Fishery Management Councils, for which the stock status is known, are overfished, and 30% are subject to overfishing. By contrast, only 13% of FSSI stocks managed by the Pacific, West Pacific, and North Pacific Management Councils are overfished, with 6% suffering from overfishing.

Only the United States has seen an improvement in performance over the past several years; in 2000, 38% of U.S. stocks for which the status was known were classified as overfished. All other areas have experienced some increases in the number of overfished stocks in the past decade, although the increases in New Zealand have been very small and are offset by recoveries in some inshore stocks. However, these statistics disguise a quite dynamic situation within each region; for example, of 74 U.S. stocks requiring rebuilding, biomass is increasing in 48% of them even if they have not yet achieved rebuilt status.

More detailed examination of the U.S. situation reveals that although ITQs are not generally applied, West Coast fisheries are managed by quota controls with fishing rights assigned to fishing companies or sectors, whereas in the northeast, fisheries are managed by a days-at-sea scheme and other effort controls. In terms of their current performance and the stock recovery required by the Magnusson-Stevens Act, West Coast management systems appear to be more effective than northeast coast systems: Only two of the 18 New England stocks that were overfished in 1995 have now recovered, compared to 4 of 9 stocks similarly categorized by the Pacific Fisheries Management Council.

Clearly, non-ITQ management systems do not always fail to maintain sustainable stocks, and management systems using ITQs are not always successful. The critical additional requirement appears to be a formally adopted management strategy with predefined rules for what to do in different circumstances. In New Zealand, in addition to an ITQ system, a formal harvest strategy embedded in the Fisheries Act (1996) means that rebuilding is statutorily required when the stock is below its target level. By contrast, the lack of a formally adopted harvest strategy in the Australian Southern and Eastern Scalefish and Shark fishery has led to an increase in the number of overfished stocks in that fishery over the past 10 years, despite operating with an ITQ system since 1992. EU fisheries also lack a formal harvest strategy; although there is a commitment under the 2002 revision of the Common Fisheries Policy to develop multiannual management plans for all stocks, such plans are only currently defined for 17 of the 94 stocks that fall under EU management, and many of these have had to be negotiated during periods of stock collapse.

The key problems, i.e., the need to provide incentives to fishers to engage constructively in fisheries management and the need to have strong legal support for predefined harvest strategies, apply equally to management of stocks under national control and those in international waters under the control of Regional Fisheries Management Organisations. To our knowledge, none of the latter currently allocate rights to individual fishers, and only a few have defined and tested effective harvest strategies. Allocation problems continue to beset these high-seas fisheries and influence compliance, data availability, and transparency.

The recovery of depleted fish stocks is a key issue and one to which most countries committed themselves in 2002 as part of the World Summit on Sustainable Development. An effective reduction in fishing effort, the participation of fishers and other stakeholders in the science and decision-making process, and the biology of the species are important factors affecting successful recovery. However, unless a harvest strategy is defined, with pre-agreed, legally binding decision rules requiring reductions of effort when stock sizes decline below limit reference points, most management authorities will still delay taking action to recover stocks. Some of this delay may arise from uncertainty in the science, but mostly it arises from an unwillingness to take decisions that will create hardship for fishers, and usually a delay will exacerbate stock decline. Ultimately, the most successful management approaches are likely to combine rights-based systems, creating incentives for fishers to operate efficiently and with long-term sustainability in mind, with a strong legal structure that requires the development of pre-agreed harvest strategies and decision rules that are triggered and adhered to as reference points are passed. As indicated earlier, an adequate control of fishing activities is also necessary.

ADDRESSING ECOSYSTEM-BASED MANAGEMENT

In recent years, there have been many calls for much wider use of Marine Protected Areas to address the need for ecosystem-based management. We see these as a useful part of fishery regulation, but they are not a universal solution because unless the basic issues of capacity, regulation, and rights are solved, protected areas will simply displace the problem elsewhere. In this we concur with recent reviews that emphasize the primary importance of conventional measures to control fishing mortality and the secondary but essential role that marine protected areas (or other area-based management, such as local prohibitions on particular gears such as bottom trawls) have in dealing with specific issues of ecosystem conservation, such as bycatch and habitat damage.

The simple creation of rights-based incentives does not automatically deal with ecosystem problems, because fishers have little incentive to minimize bycatch or habitat damage that does not affect their target species. An interesting recent development is the creation of additional incentives for fishers through market measures, such as the creation of sustainable fisheries certification schemes and pressure from environmental nongovernmental organizations for responsible fisheries. Fishers have a major incentive to improve fisheries to satisfy certification conditions, and so far most of the conditions raised in Marine Stewardship Council certifications have concerned the ecosystem effects of fishing, often related to quantifying and reducing deaths of bycatch species and damage to habitat. Even in the statistics documented for some of those states with appreciable management capacity, what is striking is for how many stocks, the status is uncertain or not determined. In the United States, the stock status of 30% of the 230 major (FSSI) stocks and stock complexes was undetermined in 2006; in Australia (48%), New Zealand (78%), and the Northeast Atlantic (61%), the numbers are even higher.

Given the problems that most authorities have in deriving reliable quantitative assessments of their stocks of major commercial importance, the large numbers of small, commercially unimportant stocks present in most areas, usually as bycatch, cannot realistically be assessed. Under a comprehensive ecosystem approach, risk assessment methodologies should be used to identify those bycatch species in need of special measures, and monitoring programs, for instance using scientific observers, need to be implemented to monitor trends in all bycatch species. The application of these approaches is in its infancy even in the most advanced management schemes; many simply respond by setting untested but hopefully precautionary effort or catch limits.

These considerations apply even more strongly to fisheries operators in developing countries. In a situation of little or no management capacity, some form of bioeconomic equilibrium is the likely result, but in such cases the management priorities may be different. Indeed, high employment with relatively modest economic rent, as long as it is compatible with the sustainability of the resource, may be a perfectly legitimate management goal. In other cases, the development of Territorial Use Rights (TURFS) within local communities can lead to effective management control and rights-based operations, resulting in successful management.

4

Integrated Fish Farming

In a number of countries in Asia (e.g., China, Nepal, Thailand, Malaysia, Indonesia) and in some parts of Africa, freshwater fish culture is integrated with the farming of crops, mainly rice, vegetables and animals (usually pigs, ducks, and chickens). This leads to greater overall efficiency of the farming system as wastes/by-products or one component are used as inputs in another. For example, poultry or pig manure can be used to fertilize the fish pond and the vegetable garden and the waste vegetables can be fed to the fish and the pigs. In Africa, fish culture in rice fields and in combination with pig and duck rearing, is not too widely practised but has significant potential. Reported fish yields ranged from 2 000-4 000 kg/ha/y with ducks, 8 500-8 900 kg/ha/y with pigs, and 3 600-4 900 kg/ha/y with poultry in Gabon. It has also been proven economically viable since it involves minimal investment. Its spread has, however, been constrained by the widespread use of pesticides in many countries (Satia, 1989).

Pen and Cage Culture

Pen and cage culture involve the rearing of fish within fixed or floating net enclosures supported by frameworks made of bamboo, wood, or metal, and set in sheltered, shallow portions of lakes, bays, rivers, and estuaries.

Compared to fish pond culture with its 4 000-year tradition, fish pen/cage culture is of more recent origin. Cage culture seems to have developed independently in at least two countries-in Kampuchea where fishermen in and around the Great Lake region would keep Clarias spp. and other commercial fishes in bamboo or rattan cages and baskets; and in Indonesia where bamboo cages have been used to grow Leptobarbus hoeveni fry as early as 1922. Since then, cage culture has spread throughout the world to more than 35 countries in Europe, Asia, Africa, and the Americas (Beveridge, 1984).

Pen culture is said to have originated in the Inland Sea area of Japan in the early 1920s (Alferez, 1977), adopted by the People's Republic of China in the

1950s for rearing carps in freshwater lakes (Beveridge, 1984), and introduced to culture milkfish in the shallow, freshwater, eutrophic Laguna de Bay in the Philippines in the 1970s (Baguilat, 1979). From there it has been successfully extended for the culture of tilapia and carps (Rabanal, 1988b). Its development and adoption as a popular technology has not been widespread, though, perhaps because of its site-specific requirements like its suitability mainly in shallow lentic environments.

At present, it is commercially practised only in the Philippines, Indonesia, and China (Beveridge, 1984). The wider popularity of cage culture as compared to pen culture may be due to its greater flexibility in terms of siting the structures.

For example, cages may be installed in bays, lagoons, straits, and open coasts as long as they are protected from strong monsoonal winds and rough seas. Floating cages can also be set up in deep lakes and reservoirs, and in rivers and canal systems, and even in deep mining pools which could not be used otherwise for culture due to harvesting difficulties (Chua, 1979 and Gargantiel, 1982).

In general, however, both pen and cage culture have expanded rapidly, especially over the past two decades vis-a-vis the decreasing availability of land-based resources for fish culture and an increasing awareness of their merits over traditional pond culture, such as:

(i) their applicability in different types of open water bodies like coastal waters, protected coves and bays, lakes, rivers, and reservoirs;
(ii) their high productivity (of as much as 10-20 times that of ponds Of comparative sizes) with minimal inputs and at lower costs to develop and operate; and
(iii) the greater socio-economic opportunities they provide to low-income families in the rural areas, particularly those displaced by the reduction of fish catches in over-exploited coastal, municipal waters, because they require comparatively low capital outlay and use simple technology.

Yields from pen and cage culture are generally high, with or without supplemental feeding depending on the natural productivity of the water body. In the Philippines, for example, the yields of milkfish from fish pens in Laguna de Bay were as high as 4 t/ha/crop (compared to a national milkfish fish pond average of 1 t/ha/y in 1980 when the productivity of the lake was very high at 1 700 mg C/m^3/hr (Baluyut, 1983).

In Indonesia, the cage culture of common carp in the Lido Reservoir in Cigombong gave a total production of 28 kg/m^2 at a stocking density of 6 kg/m^2 (Baluyut, 1983). The cage culture of marine finfishes has likewise been shown to give high yields.

Culture Species

The choice of species for stocking and rearing in pens and cages is governed by much the same criteria as in species selection for pond culture, including (Guerrero, 1982):

(i) fast growth in confinement;
(ii) good consumer acceptance;
(iii) high tolerance to a wide range of environmental conditions;
(iv) resistance to disease;
(v) ready supply of fish seed for stocking; and
(vi) ease of culture and management.

There are approximately ten species of fish which are commercially cultured in cages and pens in both temperate and tropical waters, including tilapias (S. mossambicus and S. niloticus); carps (Chinese, Indian, and common varieties); milkfish; snakeheads and catfishes; marble goby; and salmonids (rainbow trout, salmon). Marine species include mainly grouper, sea bass, mullet, snapper, and milkfish. In the Philippines, Indonesia, and China, pen culture is limited to the following species: milkfish (Chanos chanos); tilapia; and the Chinese carps: bighead (Aristichthys nobilis), silver carp (Hypophthalmichthys molitrix), grass carp (Ctenophanyngodon idella); and common carp (Cyprinus carpio). Other species have been suggested as possible candidates for utilization in pen/cage culture in the following three different environments (SEAFDEC/IDRC, 1979):

(i) Freshwater : Habitats with high natural productivity (e.g., lakes, oxbow lakes, swamps, mining pools, rivers, and reservoirs): mullets, eels, catfish, Puntius gonionotus.
Habitats with low natural productivity: Leptobarbus, Clarias batrachus, Oxyeleotris, and Macrobrachium.
(ii) Brackishwater : Sea bass, mullet, siganids, sea bream, grouper, snapper, threadfin, carangids. Hilsa spp., Sparus spp., and eels.
(iii) Marine : Siganids, pampano, yellowtail, tuna, grouper, snapper, sea bass, sea bream, carangids, pomfret.

Site Selection

The selection of sites for fish pen/cage culture should be guided by the following basic criteria (Felix, 1982; Mane, 1982; and Chua, 1979):

(i) Protection from high winds or typhoons.
(ii) Adequate water exchange that will enable the flow of nutrient-laden water through the pens/cages.
(iii) Good water quality (high or adequate dissolved oxygen, stable pH, and low turbidity, and absence of pollution).

(iv) Firm bottom mud to allow pen framework to be driven deep into substrate for better support.
(v) Freedom from predators and natural hazards.
(vi) Accessibility to sources of inputs, including labour and markets, and
(vii) Good peace and order condition.

The factors to be considered in selecting sites for pens and cages in freshwater, brackishwater, and marine environments. It is important to note that the selection of a suitable site is vital to the success of the culture system; a good site selected solves much of the management problems of pen/cage culture (Chua, 1979).

Design and Construction

Both fish pens and fish cages are built around the same basic design concept: a net enclosure supported by a rigid framework. They differ, however, in a number of respects. Firstly, a pen does not have a net bottom; the edges of its net wallings/fencings are anchored to the lake bottom/substrate by means of bamboo pegs and the lake bottom is the pen bottom. In comparison, a cage is like an inverted mosquito net with the cage bottom made of the same netting material used for its four sides.

Secondly, fish pens theoretically have no limit to their size/area while cages cannot exceed 1 000 m^2 in area for reasons of the quantity of material required for cage construction (due to the need for a flooring) and manageability of operation (cages have to be lifted and the fish scooped out and not harvested using nets as in pens).

Thirdly, design of the structures and methods of construction are different. Fish pens are fixed structures; fish cages may either be fixed or floating. Fish pens for milkfish culture in Laguna de Bay, Philippines consist of a nursery pen within the grow-out pen/enclosure. Cages are individual units for either seed production or grow-out; they are, however, usually installed in clusters or modules with a common framework. Pens and cages come in various shapes and sizes and are made of different types of materials. Most pens and cages are rectangular or square although some may be circular, as in some milkfish pens in Laguna de Bay and the milkfish broodstock cages at the SEAFDEC Aquaculture Department in the Philippines, or cylindrical as those used for fish collection in Malaysian or Indonesian fresh waters.

Rectangular cages are preferred for easy operation and management. Circular cages are more suitable for some species like milkfish and yellowtail but are more expensive to build (SEAFDEC/IDRC, 1979).

Polyethylene and nylon monofilament twine are widely used for fabricating cages and net pens although wire mesh is used in several countries. The framework structure is generally made out of bamboo and other locally available wood. Cage floatation materials include bamboo, PVC pipes/ containers, steel or plastic drums, styrofoam, and aluminum floats. The type of anchor for floating cages varies depending on the depth of water, nature of bottom, tides, and currents. Concrete slabs of different sizes and shapes, sand bags, and iron anchors are widely used in different countries.

Pen and Cage Operation

Basic procedures involved in the management of pen and cage culture are very much like those in pond culture, starting with completion of construction and preparation of the culture facilities for stocking, rearing, and harvesting. Slight variations in specific activities exist, however, as the result of the very nature of the system. For example, it is obviously not possible to apply fertilizers, lime, and pesticides since the system has open water exchange between the inner compartment and the outside environment.

Soon after construction of the pen/cage is completed, preparations are made to procure fry/fingerlings for stocking. Milkfish pens have a nursery compartment into which milkfish fry are grown for 3-4 weeks to 12 cm long fingerlings which can be released into the grow-out compartment. The nursery pen and the grow-out compartment are prepared for stocking by clearing the bottom of predatory fish like Megalops cyprinoides and Elops hawaiiensis. The milkfish fry/fingerlings from the nursery pen are stocked in the rearing pen at 20 000-50 000 per ha where they are cultured to marketable size.

In the Philippines, the milkfish stock in the pen is not generally given supplemental feeding except for occasional rations of bread crumbs, rice bran, broken ice cream cones, fish meal, and ipil-ipil leaf mill.

On the other hand, cage-reared fish may or may not be fed supplemental or artificial diets depending on the stocking density used and the level of technology in the country. Cage feeding trials in the Philippines showed the adequacy of a ration composed of 77% rice bran and 23% fish meal with feed conversion ratios of 2.2-2.8. Current feed practices in freshwater cage culture involve the provision of supplemental feeds using readily available ingredients like rice bran and poultry feeds. Other countries use artificial feeds based on simple diets preferably prepared in pelleted form for best results. At the end of the culture period, the fish are harvested from pens using harvesting nets (e.g., gill nets, cast nets, seines) or from cages by lifting the cage and causing the fish to collect in one corner for scooping out using a pail.

Open Water Culture

The farming of molluscs and seaweeds in open marine waters has become increasingly popular in a number of countries, especially in the Third World where it is seen as a viable alternative to municipal or artisanal fisheries or as a means of supplementary income for small-scale fishermen. Because seafarming is generally low-cost and labour-intensive and could thus involve entire coastal communities, it is particularly appropriate in areas where production from municipal fisheries has substantially declined and where, as a result, subsistence fishermen have little or no means of livelihood.

Mollusc Culture

Bivalves are widely cultured in a number of countries world-wide. In Asia and the Pacific, they represent a high quality food resource with annual production higher than from crustacean culture on a per hectare basis (Sitoy, 1988). In 1984, molluscs accounted for approximately 35% of the total production of coastal aquaculture in terms of gross weight in the region (Shang, 1986). The most important species for culture in Southeast Asia are the oysters (mainly Crassostrea spp.), mussels (mainly Perna spp.), clams, cockles, and scallops.

In Japan, the most commonly cultured species include Crassostrea gigas, C. rivularis, C. nippona, C. echinata, and Ostrea denseramellosa, with C. gigas as the predominant species (Honma, 1980). In Africa, the culture of Venerupis is reported in Tunisia and Pinctada spp. in Sudan (Shehedah, 1975). In Mexico, the culture of the large oyster Crassostrea spp. is carried out by cooperative societies and of the mussel Mytilus edulis on floating rafts by private investors. Oysters are widely distributed in estuaries and bays which receive some run-off from land and have somewhat lower salinity than the open sea. As they filter their food from the water, they grow best in areas with moderate to high concentrations of phytoplankton (SCSP, 1982c).

Oysters grow best in intertidal areas where they are exposed for some minutes or a few hours during low tide (Pagcatipunan, 1987). Mussels, on the other hand, cannot tolerate tidal exposure even during low tide.

The best sites for culturing molluscs are therefore those that meet their biological requirements, including the following:

(i) Seawater salinity range of 15-35 ppt.
(ii) Water depth of 1-10 m, and
(iii) Muddy bottom for mussels and hard rocky or coralline substrates for oysters.

In addition, the area for mollusc culture should be protected from strong water currents reaching three knots and should be accessible to source of seed, transport, and markets. Furthermore, the presence of local available stock in an area is a good indicator of its suitability for mollusc culture.

Countries which have successfully cultured bivalve molluscs have developed their own systems of culture which depend entirely on natural seed stock, which are either gathered from natural seed beds or collected using suitable materials for collecting seed from natural grounds (Sitoy, 1988). In the Philippines, both natural and synthetic ropes have been used for spat collection. However, since natural ropes, which have been found to attract more larvae than synthetic polyethylene or polypropylene ropes, do not last long, natural fibrous materials like coconut coir are sometimes interwoven with synthetic nylon ropes to make them more attractive to the larvae (Yap et al., 1979; Sitoy et al., 1983).

The string seed collectors are submerged in the sea water for seed collection at the right time. They are hung on a collector rack, normally 12 strings along a distance of 1.8 m to hold about 1 000 shells. Sometimes, strings are hung separately from each other at regular intervals; at others, three or four strings are put together for hanging to prevent branches from attaching to strings when they occur in large quantities.

Three principal methods of oyster culture are used in the Philippines and Japan: (i) hanging method including rafts, longlines, simple hanging, and rocks; (ii) stake or stick method; and (iii) broadcast or sowing method (SCSP, 1982c; Honma, 1980).

In Japan, the earliest method used at the Hiroshima Prefecture, where oyster culture began in the 17th century, was the stick culture method. In 1927, the hanging method of culture was introduced which later developed into different variations, viz., the simple hanging method, raft method, and longline method, to suit different local conditions as culture grounds shifted from inner to outer parts of the bay to outer open seas (Honma, 1980).

The broadcast system is actually used throughout the world in places where the bottom of shallow bays is firm enough to support the materials used as collectors and for growing oysters. Oyster shells, stones, or other hard objects are scattered on the bottom in areas where setting or the attachment of oyster larvae is known to occur. The young oysters or spat are left in places attached to the collectors until they are large enough for harvest (SCSP, 1982c).

The stake method is usually applied in shallow areas with soft or muddy bottom, usually not more than 1 m deep during low tide. The stakes, usually bamboo trunks (whole or split), branches of mangrove trees, or concrete Y-shaped posts and other similar materials are staked on the sea bottom in rows spaced about 0.5 m apart, to serve as attachment for oyster spat.

The hanging method of oyster culture uses empty oyster shells or other material such as coconut shells as collectors. The collectors are strung on synthetic twine or heavy monofilament nylon, and placed about 10 cm apart by using bamboo tubes as spacers or by tying knots in the twine. The strings are hung from a platform or rack/tray made of bamboo or wooden splits or welded wire with wooden frame, and placed on wooden plots. Oysters detached from the collectors or those small oysters/seedlings which are separated from harvested stocks are cultured on the trays until they are big enough for the market (SCSP, 1982c; Pagcatipunan, 1987).

Harvesting procedures vary with the culture method. Oysters grown on stakes or by hanging are removed from the stakes or ropes on shore or in a boat after the stakes/ropes are lifted out of the water. Those grown by broadcasting are usually collected at low tide (SCSP, 1982c).

Mussel farming makes extensive use of bamboos either as stakes or as floating rafts. The stake method, similar to that for oyster culture, is the most commonly used. The mussels are harvested by divers after 6-10 months when they reach a length of 5-8 cm.

Alternatively, mussels are grown on floating rafts which have the following advantages: (i) faster growth; (ii) possibility of regular thinning and therefore higher production per unit area; (iii) possibility of transfer to other areas to prevent siltation; and (iv) ease of construction using more durable materials (Sitoy, 1988).

Mussels and oysters grown in waters contaminated by domestic and industrial wastes need to undergo depuration or cleansing, using artificially cleaned water or clean seawater from saltwater wells, to ensure satisfactory microbiological and chemical quality of the product.

Seaweed Farming

Seaweeds, aside from being used as food, are important sources of colloids or gels, such as agar, as well as minerals of medicinal importance such as iodine. Eucheuma, a red algae, is a valuable source of carrageenan, an important industrial compound used in stabilizing and improving the quality of a great number of products. Caulerpa lentillifera, a green algae, is economically important because it is a favourite and nutritious salad dish containing essential trace minerals such as calcium, potassium, magnesium, sodium, copper, iron and zinc. It is also known for its medicinal properties, being used as an anti-fungal agent and as a natural means for lowering blood pressure. Gracilaria, another red alga, is economically important in Taiwan (PC) for its agar extracts.

The culture of the seaweed Porphyra is believed to have started as early as between 1596 and 1614 in Hiroshima Bay utilizing pole and net devices originally installed to catch fish. At present, commercial seaweed culture is limited to five countries in East Asia, viz., Japan and Korea (which both grow mainly Porphyra, Undaria and Laminaria), China (Porphyra and Laminaria), Taiwan (PC) (Gracilaria and Porphyra), and the Philippines (Eucheuma spinosium, E. cottonii and Caulerpa lentillifera). Thirty-one species belonging to 18 genera and three divisions are presently cultured in these five countries, of which only three out of the 31 species are green algae.

In 1988, the estimated world seaweed production for use in the manufacture of carrageenan was nearly 68 000 t of dried seaweeds, of which nearly 66% was supplied by the Philippines and the rest by Indonesia, Chile and Canada. The bulk of the Philippine seaweed production consists of Eucheuma produced mainly in the southern part of the country in reef-protected coastal areas. Caulerpa is also successfully farmed in seawater ponds in Mactan, Cebu (Trono, 1986).

In Taiwan (PC), Gracilaria is cultured in ponds formerly used for milkfish, with Pingtung County alone accounting for 110 ha of the total 400 ha of Gracilaria ponds in Taiwan (PC) in 1974 and producing 1 000 t of dried Gracilaria seaweed.

In Japan, indoor facilities are used to obtain buds/seedlings for on-growing at sea. The facilities consist of 70-80 cm deep square or rectangular concrete tanks provided with illumination, a temperature control system, and ventilation (Mito and Fukuhara, 1988).

The successful cultivation of seaweeds depends on four important factors (Velasco, 1988):

(i) Type of Seaweeds Used

The seaweeds cultured must be healthy and resistant to disease and breakage. They must be able to grow fast and give high yields during harvest. During processing, they must have high amounts of dry matter from which will be extracted high concentrations of carrageenan of high gel strength and viscosity.

(ii) Ecological Conditions of the Farm

The farm must be well-sited and fulfill the bio-ecological requirements of the culture species. In general, the presence of a particular seaweed species in an area is a good indicator of the suitability of that site for culture of the species under consideration.

(iii) Access to Sunlight

Seaweeds being cultivated need abundant sunlight for photosynthesis. Shading by other seaweeds and plants must be prevented by regular inspection and removal of the unwanted plants.

(iv) The Seaweed Farmer

The personality and dedication of the seaweed farmer is an important factor since the farmer must visit the farm regularly and carry out routine inspections. Some of the farmer's chores include shaking off silt and other foreign materials from the seaweeds, repairing broken lines, restoring uprooted stakes, and picking up drifting branches of seaweeds.

Trono and Ganzon-Fortes (1988) listed the following criteria for selecting good sites for Eucheuma in open waters and Caulerpa and Gracilaria in seawater ponds:

(i) Unpolluted seawater supply.
(ii) Salinity of 30-35 ppt Eucheuma and Caulerpa and 8-25 ppt for Gracilaria.
(iii) Water temperature of 27-30* C.
(iv) Moderate water movement of 20-50 m/min.
(v) Water depth of 0.5-1 m at low tides and not more than 2-3 m at high tides, and
(vi) Firm bottom protected from strong waves for Eucheuma and muddy-loam bottom for Caulerpa ponds.

Seaweeds are grown using different types of planting material (vegetative cuttings, natural seeds, hatchery-reared seeds) and methods of culture (store planting, bottom culture, rope method, rope-concrete method, and pond culture either in monoculture or polyculture with milkfish, shrimp and crabs). In the Philippines, the monoline method of culture is the most popular and successfully used of these methods (Alih, 1989). The farming activities involved in monoline culture of Eucheuma species based on the Philippine experience are as follows (Trono and Ganzon-Fortes, 1988):

(i) Securing a license from the Bureau of Fisheries and Aquatic Resources (BFAR) prior to farming the area.
(ii) Preparing required materials needed for farm construction.
(iii) Clearing the area of sea grass, seaweeds, large stones and corals, and other foreign materials, followed by measuring it according to the proposed dimensions of the farm.
Wooden stakes are then driven into the bottom with the help of an iron bar and sledgehammer and arranged into 10 m rows at 1 m intervals. An 11 m nylon line is securely tied to one end of each stake about 0.5 m above the ground and then stretched to the corresponding opposite stake and tied securely. If the current is very strong, an additional row of stakes is placed in the middle to provide additional support.
(iv) Obtaining seedlings from the nearest source and transporting them to the farm site within the shortest possible time. During transport, the seedlings are protected from exposure to sun, wind, heat or rain. If the transport of seaweeds will take several hours, the seaweeds are kept damp during the trip and upon arrival at the farm, are immediately submerged in water.
(v) Preparing the seedlings by tying bunches weighing about 50-100 g with soft 25 cm long plastic straw, and then tying these to monolines in the water at 20-25 cm intervals. The plants are allowed to grow to about 1 kg or larger before harvesting.
(vi) Building a farm house if drying of the harvested seaweeds is part of the operations. The farm house is built in or near the farm site so as not to

waste time during post-harvest handling. The size of the farm house, which is designed to provide for drying and storage, will depend on the farmer's financial capacity and market commitments.

(vii) Maintaining planted seaweeds by inspecting them regularly while they are growing. Unwanted seaweeds which will compete with the Eucheuma for nutrients and sunlight are removed along with dirt and other foreign materials clinging to the seaweeds. Lost or broken Eucheuma are replaced.

(viii) Harvesting the whole plant and reserving select portions as seedlings for the next crop.

(ix) Sun-drying of the rest of the harvest by spreading these on a drying platform of bamboo slots initially lined with coarse fine-mesh nylon net. The seaweeds are freed of all foreign matter clinging to them.

During hot and sunny weather, it takes about 3-4 days to dry the seaweeds to a moisture content of about 30% or less. The dried materials are then packed in plastic sacks for storage in a dry place or for delivery to the buyer. The pond culture of Caulerpa involves the following major steps (Trono, 1988):

Pond Construction

The pond is divided into manageable units measuring about 0.10-0.25 ha. The pond design allows for a flow-through system by providing each unit with its own supply and drainage gates. Water flows uniformly from the main gate to the secondary and exit gates during the draining and flooding process. Peripheral diversion dikes or canals along the landward edge of the pond are also built to divert run-off water from the ponds during the rainy season.

Pond Management

Water is changed daily or every other day to maintain adequate levels of nutrients. During the initial stages of growth, the seaweeds deplete the water of nutrients at a high rate and frequent water changes are needed to replenish lost nutrients and eliminate the need to fertilize. Water level is, however, carefully maintained to prevent the collapse of the dikes. Unwanted seaweeds, sea grasses, and animals which will compete with the Caulerpa for nutrients are regularly weeded out. The dikes and pond gates are inspected regularly to check for leakages, which are repaired immediately. This is vital, especially during the typhoon season.

The application of fertilizer may not be necessary as long as frequent water change is maintained. However, fertilization is resorted to when the stocks appear unhealthy and pale in colour, i.e., from light green to yellowish. When this happens, pond water is changed and fertilizer with a high nitrogen content is applied at the rate of 16 kg/ha by broadcasting or by suspending the fertilizer contained in several layers of plastic sack in strategic areas in the pond. The pond water is not changed in the next two to three days.

Harvesting

Two months after planting, the Caulerpa forms a uniform carpet on the pond bottom, a good indicator for harvest time. About 75% of the crop is harvested by uprooting the Caulerpa from the mud and placing it on to a wooden raft. About 25% of the original crop is left behind, uniformly spaced on the pond bottom to serve as seedstock for the next crop. This may be harvested after two to three weeks.

Harvested seaweeds are washed in clean sea water to remove mud and other dirt. The clean seaweeds are then placed in a basket or clean plastic sheets for further sorting and cleaning before packaging and immediate transport to the market.

Planting

To facilitate planting activities, pond water is drained to a depth of about 0.3 m. Caulerpa seedlings are obtained from the nearest source available and transported to the farm site within the shortest possible time. The ponds are stocked at a rate of 1 000 kg seedlings/ha or 100 g/m^2. A handful of seedlings is uniformly buried on one end at approximately 1 m intervals using a string as guide. After planting, the pond water is gradually raised to a depth of 0.5-0.8 m or just until the plants can be seen from the surface of the water. The newly planted seaweeds are inspected after a few days. Uprooted seaweeds are replaced and bare areas are replanted.

COMMERCIAL FISH-FARMING

The mission also visited different individuals and companies presently involved in fish-farming in the Copperbelt area and around Lusaka. They are the following:

Copperbelt are:

- Lazy A. Ranch
- W.L. Cherry
- Mrs. Rosen's farm
- Mr. Kaira's farm
- Ndola Lime Company
- Copperbelt Power Company

Lusaka area:

- Bauer farm
- Kafue Fisheries

- Mubuyu farm
- Zambia Sugar Company

Commercial fish-farmers in the Copperbelt and the Lusaka area can be grouped in two broad categories.

1. Corporate concerns who have added fish-farming to their main activities for one or many of the following reasons:
   a. as a means of providing food to their workers;
   b. to implement President Kaunda's directive to diversify in agricultural production so as to reduce Zambian economy's dependency on non-renewable resources;
   c. to recycle local funds which cannot be changed in hard currency.
2. Individual farmers who have added fish-farming to their other farming activities, as a way of using existing water resources and by-products generated by their livestock and poultry raising activities.

Main relevant characteristics of commercial fish-farming as found by the Mission were:

1. Commercial fish-farming has developed essentially in the past 5–6 years. Therefore commercial, intensive and semi-intensive fish farming for commercial purposes is a relatively new activity.
2. In general, fish-farming is not the major source of income of the concerns and individual farmers who carry it, although in some cases professional managers run the fish-farming activities on a full-time basis.
3. Fish-farming development has been financed mostly by companies and individuals own funds. Outside financing for fish farming development has been used on a limited scale.
4. Commercial fish-farmers met by the mission have shown a genuine interest and enthusiasm for their activity. Many are planning to expand their facilities. All are interested in bringing their operations to their full potential.
5. Development and management of fish-farming operations have often been carried without proper technical backing. Therefore inadequate design and pond management practices have brought insufficient yields.
6. Although many species are being raised including carps, clarias and hybrids species (O. niloticus and O. aureus) most fish-farmers have obtained best results with monoculture of O. andersonii. Former practice of raising mixed species (O. andersonii, O. macrochir, and T. rendalli) has been abandoned by many farmers.
7. Fish is usually sold at farms to traders. Farmgate prices in the Lusaka area are slightly higher, at K4.00/kg, than in the Copperbelt area where fish is sold at around K3.00/kg.

8. Different feeding and fertilizing practices are in use. Only two enterprises produce their own feed for fish. Others use chicken litter. Common practice is raising duck-cum-fish and pig-cum-fish as the main source of natural fertilizers. However, few commercial farmers keep regular records of stocking, harvesting and feeding operations which make it difficult to deduct reliable conversion ratios.
9. Fish-farmers have been encouraged by the results obtained so far and believed in the possibility of good financial returns through better management practices. Best results for pond fish-farming have been obtained at CPC farms.
10. Water supply have been a problem for some of fish farmers, mainly during the 83–84 rainy season which was exceptionally low. Increasing existing dam capacity appears to be the most convenient way to obtaining adequate water supply.

The next section gives a description of operations of different fish-farms visited by the mission.

Lazy A Ranching Co. Ltd., Chingola

Lazy A Ranch is an integrated meat producing enterprise which includes a 10,000 acres farming operation and meat processing unit with wholesaling and retailing facilities. Total investment amount to 4 million Kwacha and the turnover is 3 million Kwacha per year, mostly from cattle sales. Fish-farming operations are carried from 18 ponds of 25 × 50 m for a total area of 2.25 ha; total production is 6.3 ton per year and turnover is around K10,000. Therefore, fish farming represents less than 1% of total volume of operations. A development program is being implemented, which would create 15 additional ponds. The final objective would be to have 60 ponds of similar size (total area: 7,5 ha).

Present fish-ponds complex was developed on unused and unusable swamp land which had first to be drained; there is ample water supply from two dams developed with Lazy A ranch equipment. With 68 inches of rain the last season, compared to 56 the year before, there is no water supply problem. If need be existing dam could be raised to increase the holding capacity of reservoir.

The soil has much clay under superficial soil, which is adequate for dam and pond building. There is no seepage problem. Fish-farming development was financed by company's funds by reinvesting profits. The Shiel brothers, owners of Lazy A ranch were already involved in fish transport and distribution, distributing fish from Lake Tanganyika and Lake Mweru. They also owned fishing boats and gear operated by fishermen who sell them their catches. They knew the market and were interested in fish-farming.

Their fish-farm was started with no particular technical backing except FAO text book on Fish-farming in East Africa. The species raised are O. macrochir, O. andersonii and T. rendalli. The rendalli does not do very well but O. andersonii is giving the most interesting results.

Fish are fed leftover from feed given to cattle, including maize bran, chicken litter, and calcium phosphate. So far much of the feed seems to be lost down the pond, untouched by the fish. They are fed once a day, 5 kg/day.

Fertilizing

Pig manure carried from pig manure tank has been used as fertilizer but it does not seem to give results as good as fresh pig manure.

Price of feed

Cost of feed including cost of mixing ingredients comes to 8–13 N/kg.

Production cycle

Ponds are stocked at the beginning of winter with small fish harvested previously. They are left in the ponds for a period of 9 months. Growth is very slow during winter period. After 9 months, all ponds are reaped within a 2–3 weeks period. Then small fish fingerlings are stocked in one pond until all ponds are put under water. There has been thought of using the ponds for growing vegetable during the winter period.

Management

The son of one of the owners takes care of management.

Labour

One man is in charge of feeding the 18 ponds and maintaining them. At harvest time, one flying team of 15, part of the 140 men strong workforce, takes care of harvesting with nets. Labour costs for general labour is 2K/day.

Price of fish: K3/kg for tilapia. It has increased over the past 4–5 years while the price of beef was rather stable.

Broodstock

Initial stock was obtained from Mwekera fish farm. The best breeding period seem to be August and September.

Production

The fish-farm has been under production for three years. This year's harvest, which was done around end of the year, in a 2–3 week interval, gave an average of 350 kg/pond for a total of 18 ponds. All ponds have equal dimensions of 12.5 × 18 m^2 which is 12.5 areas. This represents a production of

6300 kg for 2.25 ha or 2.8 T/ha. They were under production for 3 years, had two harvests.

Taxation: 15% on net income for farms; 45% on net income for all other businesses.

Pond construction

The 18 12,5 × 18m^2 ponds were built in less than two months, using a D-4 and a grader. Dikes are 4 m wide at base, 3 m on top. No extra earth was needed. In the present expansion, one D-4 could clear and level 2 acres of land, an area sufficient for 3 ponds in one day.

A D-7 can build 2 ponds per day. A D-4 may be a little less. Contracted out, a D-7 costs 80K/hour and a D-4, K55/hour. Sometimes, Lazy A. ranch rents out its equipment for direct costs only (labour, fuel), that is K30/hour for the D-7.

The cost of developing a 18 ponds, 2.25 ha fish farms like Lazy A. ranch would come to K8640 not counting construction of inlets and outlets (6 days using a D-4 for clearing, 9 days of using a D-7 for pond construction). This indicate that total construction cost would be less than K10000/ha, according to the figures given by Mr. Shiel.

W.L. Cherry Ltd., Ndola

W.L. Cherry farm is located in Misundu near Ndola. It is a mixed farming operation to which fish-farming was added recently. The owner Mr. Richard Brown was interested in fish-farming because it had low capital cost, was a flexible operation, had low labour coefficient and its product is easy to market. He had 8 ponds of different dimensions, for a total of 2 ha of water. Cost of development (in 1981) was K5000/ha. The construction was done by Government Land Use Services over a period of 30 working days, at a time when rental cost for earthmoving equipment was K35/h. Commercial rate at that time was K54/h. Present rates charged by Land Use Services would now be K60/h.

However, ponds were dug in water table which makes harvesting operation difficult. During the visit to the farm, workers were ending a 2 months harvesting operation during which 60–70 kg of fish were harvested from ponds, with gill net, 4 days per week. Total harvest was about 2000 kg this year. This was the second year of operation. The first year had given a less than 1000 kg harvest. Mr. Brown was hoping for another 200 kg harvest for the next year. Large fish (200–290 g) is sold at K3/kg. Smaller fish is sold at K2.50/kg. It was observed that best growth results were obtained with O. andersonii. Poaching is a very serious problem in the area.

Mrs. Rosen's Farm, Ndola

Mrs. Rosen is a businesswoman who is fabricating and bottling a soft drink made of maize. She is considering getting into fish-farming. Next to her farm she has built a 2 ha dam recently finished. The dam was built in 6 months using only labour. A team of approximately 10 people was working 7 days/ week, using shovels and wheelbarrows. Daily rate was K3/day. "Sometimes" K2 was added for overtime. A small grader was also used for a period of 5 days for deepening the dam. The dam was approximately 100 m long, 4 m high and 1 m wide on top. Inner slope was 1/1.5, outside slope 1/1.

Thus it is estimated that the whole job represented a total volume of earthwork of 2400 m^3. Each man-day represented approximately 1.3 m^3. Counting only cost of regular time (K3/day), cost per m^3 of earthwork would come to K2.28.

Mr. Kaira's Farm

Mr. Kaira heads a transportation company. He is doing fish-farming with other farming activities on a part-time basis near Kitwe. He developed his farm through the year 82, 83 and 84, using only labour.

The fish-farm now includes 3 ponds of 40 × 26 m, 5 of 72 × 18 m, 1 of 60 × 28 m and one breeding pond of 5 × 15 m. Ponds are about 1 m deep with 0.5 m freeway. One pond was partially dug in the rock. Broodstock was obtained from Mwekera fish-farming station. Stocking has been done in 1983 and 1984 but he did not know when was the right time to harvest. On the day the mission visited his farm, it was obvious that some ponds were full of fish. Fish are fed chicken manure bought from Canso Farm at K1/bag of 60–70 kg. It was specified that this chicken manure is not the one made with saw-dust, which is usually kept for gardening. Other farms sell their chicken manure reportedly at K9/kg of 60–70 kg.

Ndola Lime Company, Ndola

Ndola Lime Company is one of the major mining companies in Zambia. In 1979, the company built 3 fish ponds on an experimental basis, as a part of its plan to diversify into agricultural activities. Each pond is approximately 70 × 25 m. Dikes are approximately 6 m wide. All these ponds were built with a company-owned D-8, part time on a three-month period in 1979. Production started in 1980.

There are plans to expand to a total of 15 ponds of similar size. Water is supplied from an old company quarry used as reservoir. Initial stock of O. macrochir, O. andersonii and T. rendalli was obtained from Mwekera fish farm. Production is bisex. Ponds are fertilized with duck manure (1800 ducks are kept in a duck house near pond N°1). Chicken manure is also used at the rate of 50 kg/pond/week. Also feed judged unsuitable for chicken is put in ponds. Inorganic fertilizer (superphosphate) is also put in ponds as supplement fertilizer at the rate of 10 kg/month. Production cycles last 12 months. One partial crop in 1982 from the three ponds gave a gross production of 1695 kg. Another crop brought 1350 kg in May 1983.

Fingerlings are restocked in ponds. Market size fish is sold to company employees at K2,0/kg (1984), alongside other products from company's farm. Chicken are sold at K5/head (N1, 5 kg), ducks are sold at K7,5/head (around K2/kg).

Copperbelt Power Co. Ltd., Kitwe

C.P.C. farm was developed in 1979 (1st phase, 14 ponds), following a president's request that companies start farming. Total pond area is now 7.4 ha. Production of fish started in 1980. Initial stock was provided by Mwekera Station: O. andersonii, O. macrochir, T. rendalli. Best production results were obtained with that O. andersonii. Introduction of livestock came later around 1981. Ducks were introduced for getting manure for the fish. Pigs were introduced in 1984. Dairy cattle (Jersey cows) were introduced in 1983, starting with 14 heads.

C.P.C. has been raising carp, at first mixed with other species, now in a separate pond. The stocking was done in November and harvesting has yet to be done.

All ponds are now stocked with three spotted bream (O. andersonii) except for pond #28 which is used as feeding pond for the yellow belly fish (Seranochromis robustus) which will be stocked as predator and pond #6 which is used for producing carp.

Feeding

The feed consists of chicken litter sweepings containing 20% of maize bran. Feeding is done twice a week, at a rate of 130 gr/m^2 week but actual feeding would be closer to 100 gr/m^2 week. Chicken litter is a by-product of the chicken raising operation. CPC has 43 chicken houses, with 1000 chicken each. Each house is swept every three weeks, giving 50–60 bags. Each bag officially weighs 50 Kg but the real weight is closer to 40 kg (38T of manure/1000 chicken/yr). Chicken litter is also sold out at a price of 3K/bag (if bag is provided) or 4K/bag. Two thirds of chicken manure production is sold to the public.

Fertilizing

Some inorganic fertilizer (2C phosphate) was used at the beginning of operations but the practice was abandoned. Now, pig manure is used for one pond. Pig manure production was 400 kg last year: maximum production would be 600 kg for the existing number of pigs. No chicken litter sweeping is used in combination with pig manure.

Duck manure is also used but ducks are not kept on the water since it causes production losses.

Yields

For the past year, maximum yield was 8,7 T/ha/yr, lowest was 1.3 T/ ha/yr, and average was 4.07 T (net production of ponds in production). All those performances were obtained from stocking with mixed species. Pond with highest performances is pond N°17, a small pond of 1599 m^2, maximum depth of 1,62 m and minimum depth of 0.4 m.

Labour

Operations are done by 8 laborers paid 67 N/hour for a 45 hours week (5½ days) and one supervisor paid at 90 N/hour for the same workweek. (Total labour cost: K15000) Main activities are feeding, harvesting and net mending. Extra labour is occasionally required for repair work and slashing grass.

Management

Fish-farming operation management is done by 2 managers who are sharing their time between fish-farming and other farming activities, and one full time advisor-biologist.

Production

Total gross production for 84–85 period (April 84 – March 85) was 32.98 tons. Of this total, 6.5 tons were kept for restocking. Net production is estimated at 26. Production for 83–84 was 18 T. Production for 82–83 was 29 T. For those three years, the same pond area was under production. Low production of 83–84 was due to the fact that ponds were cropped only once. There was less management input during that year.

Production cycle

Production ponds are stocked twice a year. So far ponds 10 and 11 were used as breeding ponds but were cropped only twice a year. CPC had 10 crops in March. On average, there is a crop once a week.

Sales

Total sales for 84–85 period were 52,000 K, that is 26,000 kg sold at an average price 2.0 K/kg.

Other stocks: Pigs: 51, Chicken: 43,000, Ducks: 290

Other production: Pigs: 200, Chicken: 120,000, Ducks: 5,000

Bauer Farm, Lusaka

Bauer Farm is located near Lusaka. The main source of income is a meat processing unit of 70 pigs/week capacity. The farm has two production ponds of 20 × 80 m each. It has also a breeding pond of 15 × 25 m^2 (depth: 1.5 m). The production ponds were built with Mr. Bauer's own staff and equipment (caterpilar) 3–4 years ago, the cost of fuel was K3200. The feeding pond was built at a cost of K1200 (K300 of earth movement and K900 of labour and bricks).

The water has to be pumped from the river and water requirements are high since the soil is sandy, causing much seepage. Bauer Farm raises carps obtained from South Africa and from Chilanga. Carps breed once they have reached approximately one kg. Production ponds are harvested once a year at Christmas time. Total harvest is 500–600 kg, representing a gross production of 3,125 T/ha/yr to 3.75 T/ha/yr. The small fish is given to the staff. Individuals of more than 1.5 kg are sold mostly to resident Russians and Yugoslavs, at K8–9/kg.

Despite the problems created by sandy soil, Mr. Bauer would be interested in expanding pond area to 10 ha, provided he can find a capable manager.

Kafue Fisheries, Kafue

Kafue fisheries is located at 90 km from Lusaka, on the Kafue river. It has 18 ha of ponds and is owned by T.A.P. building products (managers: Mr. and Ms Flynn). Development of the fish farm was as follows:

October 81: 5 ha of production ponds and 0.23 ha of feeding ponds plus road.

October 82: 5 ha of production ponds.

1983: No development.

1984: Development of 7.5 ha of production ponds.

They do not want to grow bigger because they could not grow more pigs. They are thinking of expanding in rice, as the government is expected to ban rice imports in 1987.

Water is supplied from the Kaufe river which flows alongside the farm. All the water is being pumped.

Temperature

Elevation being approximately 1000 feet lower Lusaka, average temperatures are higher than at Chilanga Station.

Associated husbandry

600 pigs are kept on the fish-farm. Each production pond has its pigsty. The ratio is 50 pigs/ha. Pigs are purchased as weaners at 50–60 kg and sold at

90–100 kg. It sells 2000 pigs per year. Main species raised is Oreochromis andersonii. Other species were raised previously, like the oreochromis mossambicus, but the focus is on O. andersonii which gives the best yields. Also, Seranochromis robustus (yellow belly fish) have been purchased from local fishermen at market price to act as predators. The stocking process includes 3–4% predators. The stocking density is 30000/ha, the weight of fish varying between 10–20 grams. Use of predators has increased from 70% to 90–95% the percentage of marketable fish.

One pond has been stocked with carp but the results have not been satisfactory because of low stocking density (the carps were bought from Chilanga fish farm).

Yields

Yields are superior to 4 T/ha/yr. Their objective is to rise them to 6 T/ha/yr. Fish which are too small are dried on concrete slabs and sold as feed to Zambia Sugar Company.

Feed

Only the pigs are fed, 3 times per day. The pigsty is cleaned twice a day. Pigmeal is purchased from National Milling Co. Occasionally, yeast is added to pigmeal, as well as some fish meal. Stockfeed for pigs amount to 8 tons/week. Cost is (K30/kg).

Water needs

The fish farm requires 39,000 m^3/ha/yr of water. Diesel costs K47,000/yr, which is 40% for lighting, 40% for pumping and 20% for boreholes. Pumping is important due to a high degree of evaporation.

Construction costs

Cost of pond construction since the beginning of the farm comes to K10,600/ha of pond. All asbestos built pigsties have been built at a cost of K1300/unit.

Annual turnover for Kafue Fisheries is K600,000. Operation costs are K300,000/yr, including K36,000 for labour (27 laborers). Depreciation is approximately K100,000 which makes the operation seem profitable for a farm whose development cost K900,000.

Mubuyu Fish-Farm

Mubuyu farm covers 3000 acres used to grow wheat of 25 ha of reservoir which are filled half of the year. Water required for growing wheat has to be stored in those reservoir for irrigation purposes. At the moment, water used for growing wheat brings better financial returns than water used for growing fish.

The farm has 50 ha under water, including 15 reservoirs covering 25 ha. The hatchery complex covers 4 ha, including 10 circular tanks for larvae rearing. It has an incubating capacity of 1,000,000 eggs.

Ponds were constructed in 1981. Construction was contracted out to a firm. Total cost was 0.5 million K. Unit cost was $3K/m^3$.

Species

Production was started with imported species. Local species were added: andersonii, macrochir, mossambica, rendalli. They were first put in spawning ponds and then rearing ponds.Those species have given very small fish. Imported specimen from Israël, Malawi and Zimbabwe are grass carp, silver carp, mirror carp, nilotica and aurea. These were stocked in April but are still in breeding ponds.

Artificial breeding

Artificial breeding is carried according to instructions given in Woynarovitch's book, by hormone inducement.

Production cycles

Hatchery: 10 days.

Juvenile rearing tanks and ponds: 3 weeks

Production ponds: rest of the year (8 months, not counting winter months).

Production results

Since 1981, a total of 120 T of fish was harvested which averages 30 T/year. There is no figure on comparative yields of different techniques. Artificial and organic fertilizing, crop waste, soya bean, wheat and sorghum are in use. The farm has equipment for making feed pellets.

Stocking density: $2/m^2$: Losses are estimated at 50%, mostly by theft; predators are also active. There is no sexing of fish.

Number of employees: 20, for fish farm only.

Payroll: K1500/month, not counting management time.

Power costs: Electricity bill comes to K1800/month (fish farm and hatchery).

Transportation

Feeding is done by using 6 oxcarts, 12 oxen. One cart lasts 5 years, costs more than K1000 and carries 500 kg.

Sales

Fish is sold at the farm, to traders, at 4K/kg. Catfish is filleted and sold at K8/kg. Catfish acts as predators. At harvest time, reservoirs are emptied. A sample of 20 kg of fish has been sent to Japan and an order is expected.

Under development

A 300 acres reservoir with 16,000,000 m^3 holding capacity will be ready soon. The dam will have 440,000 m^3 of earth. Cost of construction will be K6/ m^3 using mostly own equipment. Market rate would rather be K20/m^3. The reservoir would cover two years of water needs (annual needs for 1,200 ha being 8,000,000–9,000,000 m^3).

Zambia Sugar Company (ZSC), Mazabuka

An extensive fish-farm with a capacity of 1 ton/week of fish, has been developed downstream from ZSC network of 11 reservoirs and upstream from its 11,000 ha of irrigated sugar cane fields. Therefore, the fish farm is not costing the company any money in terms of water use. The main purpose of the fish farm is to supply the company's kitchen with a regular supply of fish for its workers. Relative price of fish in comparison to beef has been decreasing, as the price of beef was going up. Also, coming from the western area of the country, workers have a preference for fish over beef.

Water supply

Water supply comes from a network of 11 reservoirs used to irrigate 1100 ha of sugar cane field. All water has to be pumped from nearby Kafue river. The fish farm being located downstream from those reservoirs and stream from cane fields, water used by the fish farm is recycled in the irrigation scheme. Therefore the only additional cost for water is the extra length in pipes.

Species of fish raised

Production is concentrated on hybrid obtained by crossing tilapia aurea male with tilapia nilotica female taken from an original stock which was purchased from Sterling University.

Production cycles

Ponds	Duration	Final weight	Size of Pond	N.
1) Breeding ponds	2–3 days	<1 GR	177 m^2 (15 m. diam.)	1
2) Raceways	3 weeks	1 GR	10m×75(m×0.25)	11
3) Fry tanks	3 weeks	3–5 GR		6
4) Fingerling tanks	6 weeks	10 GR		
5) Rearing units	20 weeks	250 GR	10×1.5m=11m^3	20
Total	32 weeks			

Breeding ponds

Ponds with mud bottoms are used to produce straight aureus, nilotica and andersonii. They are six in number, dimension 12 × 25 m (300 m^2). They are stocked at a rate of 200 female and 40 male per pond. Production is 50 fry/

female/month from September to end of April. The round arena pond with fingerling traps and ring for females has 19 m diameter and 1 m depth; its area is 177 m^2. It produces 10,000 fingerlings per week of macrochir. Production of hybrids is more difficult. The 2 other concrete ponds have 10 × 25 m^2. They produce 20,000 – 25,000 fingerlings/year. Mud bottom ponds have the disadvantage that fingerlings tend to hide in nests at time of harvest. Therefore, concrete bottom ponds are preferred. The production of fingerlings was 250,000 for the 84–85 season, not counting andersonii.

Resting ponds

Four circular of 6–7 m diameter and 0.75 in depth are used for holding the original stock received from Sterling University, and also for holding crayfish. Trial ponds: ZSC has 10 trial ponds, of 3 m × 0.75 m each, to try new feed, new vitamins.

Rearing tanks

20 rearing tanks of 10 m × 1,5 m × 1 m depth with a baffle in the middle are used for bringing fish from 10 grams to 250 grams in a period of 20 weeks. Each tank holds 5000–6000 fish. A harvest time, some 4000 are left, as during growth period, bigger fish are stocked in tanks stocked earlier and smaller fish are put in a dam outside. At harvest time, water level is lowered and fish are netted out. Each tank uses 8 m^3 of water per hour. The whole system (hatchery and production) uses 200 m^3/hour of water all year round. There is no need to pump the water (that is for fish farming). There has been no water supply problem so far.

5

Fish Production

RECENT TRENDS IN FISHERIES PRODUCTION.

The Global Position

The first comprehensive estimate of global fisheries production, produced by FAO in 1945, suggested that the total marine harvest was probably in the region of 17.7 million tonnes, with the majority (circa 95%) being landed by commercial fleets. These fleets moreover concentrated their activities almost exclusively in the North Atlantic (46% of marine landings) and North Pacific (47%) regions and so, while experts recognized that some European stocks were already over-fished, there was a prevailing mood of optimism in the immediate post-war period about the potential for fisheries expansion. Under-exploited stocks off Central America, Peru and Chile, in the Caribbean, off West Africa and off Australia, New Zealand, the South Pacific Islands and the East Indies in particular, were viewed as offering considerable possibilities for fisheries expansion.

Fifty years on, perceptions had changed. No longer was the talk of fisheries potential and abundance, but of over-exploited stocks and excess fishing capacity. Increased investment in fishing activity, firstly by developed nations—often in the guise of distant water fleets—and latterly by developing nations, saw impressive fleet and employment growth over the period.

Although FAO warn of data shortcomings vis-à-vis employment within the sector, it is undeniable that the number of fishers—whether full or part-time or occasional—has climbed dramatically in the developing countries. In Asia, for example, the growth in aquacultural activities and increased marine and inland capture sees the fisheries sector now employ over 32 million people (87% of the global total). Meanwhile, although the numbers employed have decreased in the fully developed, largely industrial fisheries of Europe and North America, this decline has been more than matched by the growing popularity of recreational fishing in these regions. There are sharp regional differences in labour productivity, however. In 1995, for example, 301 000 Japanese fishers produced 6.7 million tonnes of fish—while nearly 6 million Indian fishers produced around 5 million tonnes of fish. This global increase in employment has been accompanied by increased fleet investment. Data, available on the global level since 1970, suggests the global fleet of decked vessels rose from 595 099 (with a capacity of 13.4 million tonnes) in 1970 to 1.26 million vessels (capacity 28 million tonnes) by 1995. In addition, estimates suggest a further 2.8 million undecked vessels are presently deployed in fishing activities, with the major proportion of both decked and undecked vessels (83 to 85%) being located in Asian waters.

This investment in terms of both money and time, in turn has fed through into substantial production increases in both marine capture fisheries, and latterly aquacultural (marine and inland) output.

In the case of marine capture fisheries, catches rose sharply from the 1950s through to the late 1980s—since when production has stabilized. Over the same period, aggregate pelagic landings have increased from 8.1 to just over 40 million tonnes (47% of marine catch), albeit with rather large oscillations due to climate-induced variations in natural productivity and the effects of boom/bust fishing strategies. The exponential growth in developing countries'—principally Peruvian—take from the late 1950s was stymied by the anchoveta collapse in the early 1970s, although by the end of the century such countries accounted for over two-thirds of the global pelagic catch (a major proportion of which is destined for the reduction industry). While around 200 pelagic species are commercially fished, seven species (*anchoveta, Atlantic herring,* the *Japanese* and *South American pilchard,* chub mackerel, *capelin* and the *Chilean jack mackerel*) account for the bulk of the landings and of these six (in italics above) are presently considered to be fully exploited or worse (overexploited, depleted or recovering).

The remainder of the marine catch is comprised of demersal fish (15% in 2004), molluscs (8%) and crustaceans (7%)—which are discussed more fully in the relevant regional reviews that follow, unidentified species (14%), miscellaneous coastal fish (7%) and sharks, rays and diadromous species (2%).

The aggregate landings of demersal fish have risen to a current peak of around 22 million tonnes, with developing countries now accounting for half the reported catch. In contrast, the rather more dramatic increases in developed country catches during the 1950s and 1960s triggered the introduction of fishing restrictions in Icelandic waters in the early 1970s and then the collapse of ground fish stocks off eastern Canada in the early 1980s. Since then, landings by developed country fleets have fallen by around 30%.

Nevertheless, the most dramatic growth over the last half century has been seen in the fields of marine and inland aquaculture. An industry producing 638 600 tonnes in 1950 has evolved into a 59.4 million tonne operation generating US$70.3 billion by 2004.

While output in the developing world has leapt twelve-fold to 4.4 million tonnes, this increase has been completely outshone by the two hundred-fold increase in production across the developing world, most notably in Asia. Individual regional trends in aquacultural activity—and the evolution of marine and inland fisheries—are discussed in more detail in the regional reviews that follow.

To conclude this section, however, we wish to draw attention to the fact that this expansion of fishing effort, fishing employment and fish production has not been without cost. The most recent stock status research suggests that over three-quarters of the main commercial stocks are presently fully-overexploited (or worse). It seems clear then that prospects for future fisheries growth are limited.

THE REGIONAL PERSPECTIVE

Asia

Asia is the world's foremost capture fishery and aquaculture producer. In 2004 the regional marine harvest alone totaled 32 million tonnes of marine fish, 16 million tonnes of molluscs, 14.4 million tonnes of aquatic plants and 5.9 million tonnes of crustaceans.

Although the growth in marine fish landings since the early 1990s has been rather modest (up 23% from around 26 million tonnes) due to the gradual (over) exploitation of stocks, crustacean extraction has doubled and the harvesting of molluscs and aquatic plants has trebled over the same period. The principal marine fishing nation in the region is China, although a further nine states landed in excess of 500 000 tonnes in 2004.

The Table, nevertheless, fails to illustrate the profound re-alignment in regional production that has taken place since 1990. In that year Japan landed 8.1 million tonnes (30.7% of regional landings) and the Republic of Korea 1.8 million tonnes (6.8%). However, a series of fishing capacity management plans introduced in the two countries during the intervening years has seen capacity—and with it marine landings—sharply reduce in both countries. In contrast, production has soared in most of the less developed economies (Chinese landings up 147%, Vietnamese up 132%, Indonesian +92%, Myanmar +84%), India being somewhat of an exception (production up 23%) due to its principal marine fisheries already being rather closer to full exploitation.

Country	*Million tonnes*	*Percent of Asian landings (Cum. percentage)*
China	10,364,606	32.0 (32.0)
Indonesia	3,840,189	11.9 (43.9)
Japan	3,345,378	10.3 (54.2)
India	2,340,189	7.2 (61.4)
Thailand	2,269,521	7.0 (68.4)
Philippines	1,924,715	5.9 (74.3)
Vietnam	1,333,811	4.1 (78.4)
Rep. of Korea	1,130,736	3.5 (81.9)
Myanmar	1,091,740	3.4 (85.3)
Turkey	502,312	1.6 (86.9)

Table : Major Asian marine fishing nations (>500 000 tonnes), 2004.

The principal molluscs extracted are the high-value Pacific cupped oyster-*Crassostrea gigas* (4.3 million tonnes in 2004) and Yesso scallop-*Patinopecten yessoensis* (1.5 million tonnes), and the lower value Japanese carpet shell-*Ruditapes philippinarum* (2.9 million tonnes). While China dominates mollusc production in the region (61% of reported 2004 regional harvest), Japan (1.8 million tonnes, 11% of regional harvest), the Philippines (1.3 million tonnes, 8%) and the Republic of Korea (800 000 tonnes, 5%) also possess significant mollusc-extraction sectors. The main aquatic plant harvested is Japanese kelp, the harvest of which doubled to just over 4.6 million tonnes during the period 1990-2004, although regional production of *Wakame* (a kelp-like plant), laver (*Nori*) and Zanzibar weed have also surpassed the million tonne mark (2.5 million, 1.4 million and 1.1 million tonnes respectively in 2004). The main driver in aquatic plant extraction is, once again, China (76% of 2004 regional harvest). Although prawns/shrimp dominate under the crustacean category (73% of category volume in 2004), serious viral pathogens affecting the region's main indigenous *penaeids* (the Giant tiger prawn and the fleshy prawn) constrained growth for much of the 1990s. The introduction of the whiteleg shrimp at the turn of the century however promptly restored momentum to the sector, with this one species alone providing a harvest of 710 465 tonnes (17% of regional shrimp/prawn harvest) in 2004. China supplied 53.5% of the regional crustacean catch the same year.

Regional production from inland water fisheries has grown more than three-fold since 1990 to 27.4 million tonnes in 2004, although the majority of this (86.4%) arises through aquacultural operations. China dominates once more (67.8% of 2004 harvest), having increased landings from 4.9 to 18.6 million tonnes over the same period. India (2.95 million tonnes), Bangladesh (1.5 million tonnes), Indonesia and Myanmar (824 000 tonnes apiece) and Thailand (531

000 tonnes) also post significant inland landings. Carp—chiefly Common, Crucian, Grass and Silver—account for two-thirds of the regional inland fisheries output. FAO calculations suggest that aquaculture production provided an income in 2004 of around US$27.9 billion for participants, up from US$7.3 billion in 1990—with the main beneficiaries currently being China (US$23 billion) and India (US$2.2 billion). Per caput food supply varies across the region being highest in East and Southeast Asia (26.8 kg), and markedly lower in both Southern Asia and the Near East (both 5.6 kg). Per caput supply is particularly high in Japan (64.7 kg), where the tradition of eating fish is very strong and fish is generally more important than meat in the diet, while the converse holds true in Afghanistan (zero kg). In Bangladesh, Indonesia, and Sri Lanka fish provides more than half of the daily animal protein requirements.

Europe

In Europe (including the former USSR and the new transition economies) marine catches have shown a sharp decline from the 20 million tonnes that was regularly landed in the 1980s, with the most recent harvest data (2004—12.6 million tonnes) being on a par with the quantities harvested in the early 1960s. The major contributor to this decline has been the collapse of the USSR—with Soviet landings dropping from 11 million tonnes in 1987 to just 2.5 million tonnes in 2004. In Western Europe, the intensive exploitation of many demersal groundfish stocks (cod and haddock, for instance) during recent decades has meant that some stocks are now considered to be outside the safe biological limits and, as a consequence, European Union (EU) fleets are increasingly seeking access to other countries' exclusive economic zones (EEZs) for their distant-water fleets. The major (over 0.5 million tonnes) European marine fishing nations in 2004 were the Russian Federation, Norway, Iceland, Denmark, Spain, the Faeroe Islands and the United Kingdom.

Table : Major European marine fishing nations, 2004.

Country	*Million tonnes*	*Percent of European landings (cumulative percentage)*
Russian Federation	2,504,428	20.1 (20.1)
Norway	2,455,439	19.7 (39.8)
Iceland	1,695,103	13.6 (53.4)
Denmark	965,942	7.8 (61.2)
Spain	739,416	5.9 (67.1)
Faeroe Islands	585,835	4.7 (71.8)
United Kingdom	522,063	4.2 (76.0)

A similar trend is evident in inland waters. Here catches have fallen from a late 1980s peak of 450 000 tonnes to 206 062 tonnes in 2004, the principal factor in the decline being a sharp fall in reported landings from the ex-Soviet economies. The one success story has been aquaculture. In 2004, 2.24 million tonnes were produced at a value exceeding US$5.58 billion—up over 50% in value and volume terms since 1994—with farmed diadromous species (principally salmon and trout) accounting for the bulk of the growth and the major portion of current worth (57.5% of 2004 revenues). The main producing countries were Norway and, to a lesser extent, Scotland.

While current fish per caput supply levels in the region as a whole (at around 17 kg) are slightly above the global average, levels vary considerably among sub-regions and countries. While in the industrialized countries the average per caput supply is about 22.5 kg per year, in Iceland mean fish supply exceeds 90 kg per capita. In contrast, in Bulgaria, Rumania and Serbia-Montenegro, annual supply is less than 4 kg per capita.

Latin America and the Caribbean. In Latin America and the Caribbean, total marine landings are heavily influenced by fluctuations in small pelagic (principally anchoveta) stocks, and periodic stock collapses (such as the anchoveta collapses in 1973 and, to a lesser extent, in the early 1980s—and subsequently in 1998) result in sharply reduced harvests. This has been slightly offset by an increase in demersal production (particularly hake) and squid in the southwest Atlantic and tuna across the region, but most major demersal and small pelagic stocks are now considered to be fully-or over-exploited. Marine production in the region peaked in 1994 when 22.8 million tonnes was landed, with current (2004) landings around 17.5 million tonnes. Major Latin American marine fishing nations (over 0.5 million tonnes) in 2004 were Peru, Chile, Mexico and Argentina.

Table : Major Latin American Marine Fishing Nations, 2004.

Country	*Million tonnes*	*Percent of Latin American landings (cumulative percentage)*
Peru	9,239,150	52.7 (52.7)
Chile	4,615,458	26.3 (79.0)
Mexico	1,079,337	6.2 (85.2)
Argentina	761,447	4.3 (89.5)

In comparison, Latin American inland fisheries are rather less significant, yielding around 325 000 to 375 000 tonnes per annum over the last few decades, with the recent rise in recorded landings (to around 440 000 tonnes) more due to better recording of catches than the exploitation of new fisheries. The same cannot be said of aquaculture, however. The abundance of small pelagic fish provided the stimulus for the development of an important fish reduction industry (currently consuming more than two-thirds of the total marine catch) and with it, a domestic aquaculture industry. The contribution of aquaculture—

in volume and value terms—increased by more than threefold over the decade to 2004, production increasing from 0.34 to 1.14 million tonnes while the value of aquacultural output grew from US$1.32 to 4.56 billion. The main impulse for this growth came from the expansion of freshwater fish and marine diadromous (essentially salmon and trout) farming—up five and six-fold respectively, with Chile exclusively responsible for the upsurge in diadromous output and Brazil largely accounting for the increase in freshwater fish landings.

As in other regions, fish supply and consumption levels vary widely. In some of the small Caribbean island states (Grenada, Montserrat, the Turks and Caicos islands) per caput supply exceeds 40 kg p.a. —substantially higher than in either Central America (around 9.3 kg p.a.) or South America (8.8 kg p.a.)—while its contribution to daily animal protein requirements ranges from 1.7% in Bolivia to 60.5% in Belize.

Africa

African marine catches, having risen from 3 million tonnes in 1990, stabilised at around 4.8 million tonnes at the start of the twenty-first century. However, these aggregate figures mask a number of differing trends. Landings in North Africa, which accounted for around 26% of the 2004 catch, are dominated by Morocco (67% of sub-regional landings), which realizes 95% of its catch in the Atlantic. Contemporary evidence suggests there is little scope for further exploitation of Mediterranean fish stocks. Southern Africa harvests have grown above trend, enabling the region to increase its share of continental landings from 30 to 34% over the period 1990-2004. Production is dominated by South Africa (52% of sub-regional landings), with its focus on the Cape hake fisheries, and Namibia (34%), where the emphasis is on small pelagic stocks. Eastern sea-board landings are much less important, accounting for under 4% of continental landings in 2004, a percentage that has shrunk moreover as landings in the sub-region have risen at well below trend rates since 1990. The Western sea-board supplied 36% of the marine catch in 2004, largely drawn from Senegal (374 245 tonnes in 2004), Ghana (313 935 tonnes) and Nigeria (251 232 tonnes). Although the Ghanaian catch in 2004 is of a similar magnitude to that recorded in 1990, Senegalese pelagic landings have increased dramatically (100 000+ tonnes) over the same time-span.

Table : Major African marine fishing nations (>500,000 tonnes), 2004.

Country	*Million Tonnes*	*Percent of African*
South Africa	866,928	18.1 (18.1)
Morocco	854,298	17.8 (35.9)
Namibia	564,819	11.8 (47.7)

Africa has the world's most important inland fisheries, 2.25 million tonnes being harvested in 2004 compared to 1.8 million tonnes a decade previously. Here too though the aggregate figure obscures regional variations. Tanzania and Kenya, who source a large portion of their inland catch from the Lake Victoria Nile perch fishery have experienced landing declines of 16 and 38% respectively since 1990 (although Ugandan catches have remained rather more buoyant). Elsewhere, however, the story has been one of increased exploitation of inland water resources—catch estimates for the Democratic Republic of Congo suggest landings now stand at 217 000 tonnes (up 35.6% since 1990), Egypt has seen output leap by 52% (to 238,455 tonnes), while Nigerian landings have doubled to 215 000 tonnes.

Aquaculture, by contrast, is less well-developed in the region, output reaching 570 113 tonnes, worth US$893 million in 2004. Nevertheless, this represents a substantive improvement when set beside the equivalent 1990 figures of 82 475 tonnes and US$166 million. Egypt is the principal aquaculture producer, presently accounting for around 83% of volume and 69% of value, with common carp culture well to the fore. Other countries with briskly growing aquaculture sectors include Kenya, Madagascar, Nigeria, South Africa and Zambia, all of whom have doubled their annual production (admittedly, from a small initial base) several times over the last decade or so.

Per caput fish supply in the region is low by international standards. Sub-regional averages lie between 3.7 kg (East Africa) and 11.6 kg (West Africa) p.a., with Niger (1.6 kg) and Gambia (29.3 kg) representing the extremes. Fish, historically, has nonetheless been a popular food item, providing more than half an individual's daily animal protein needs in Gambia, Ghana and Sierra Leone and—since most parts of the fish are consumed—it has also contributed significantly to calcium and iodine intakes. However, the growth of the fish trade with the EU (facilitated, in part, by fishing agreements between the EU and various African nation states), Japan and, more recently, China, is leading to a reduction in the domestic availability of fish in both the main West African nations and those landlocked Central African countries that used to import fish from West Africa.

North America

In North America, marine catches declined steadily from a peak of 6.8 million tonnes in 1988 to 5.6 million tonnes in 2004, principally due to the overexploitation of the main commercial groundfish stocks in some areas where fisheries are now closed or subject to restrictions. The most notable closure was in 1992 when Canada closed its northern cod fisheries off the Grand Banks, and over 20,000 people lost their jobs. Canadian landings, as a consequence, have fallen rather more sharply (down just over 50% since 1990, to 614 107 tonnes) than US landings over the last decade or so. A similar trend is evident in inland waters, the take dropping from a 1989 peak of 205 503 tonnes to the current level of 143 370 tonnes (down 30%), despite the growing popularity of sport fishing, most notably in USA where federal agencies have been particularly active in developing recreational fisheries programmes since 1995. In the aquaculture sector, however, production in both volume and value terms has more than doubled since the early 1990s (currently output is 955 178 tonnes worth US$1.99 billion). Aquaculture in North America—unlike in Europe or South America—is a much more diversified industry which includes marine, freshwater and diadromous fishes, crustaceans, molluscs and plants, although salmon and trout dominate in Canada, while catfish is an important sub-component of the US industry. Per caput fish supply levels in North America—at 22.7 kg per head—are comparable to the levels recorded in the industrialized countries of Europe, and sharply higher than the rates posted in South and Central America.

Oceania

Oceania provides comparatively little of the global marine catch (<2%), landing 1.4 million tonnes in 2004 although this is certainly an underestimate as much of the coastal catch in the region's small island developing states (SIDS) goes unreported. Although catches have risen by around 53% over the past decade, concerns over the current status of major commercial stocks has prompted the introduction of individual transferable quotas (ITQs) in New Zealand, the main regional fishery. Besides New Zealand (44% of regional landings—cod and molluscs each accounting for a third of the NZ catch), the other major regional fishing nations are Australia (18%, a quarter of which comes from the crustacean catch) and Papua New Guinea (15%, almost exclusively tuna). However, for many SIDS it is a key strategic sector in the quest for development given its contribution to production, local food security, employment and export earnings.

Landings from inland waters are equally low in global terms, with a reported catch of just 11,313 tonnes in 2004—largely from Papua New Guinea (96% of total), where inland fisheries in the highland areas are an important source of food security. However, regional data does not presently include sport and recreational fisheries—important fisheries in both New Zealand and Australia. Oceania's contribution to global aquaculture output is also relatively small, amounting to 139,273 tonnes worth US$447 million in 2004, although it has grown by 86% and 166% in volume and value terms respectively over the past decade. Production is almost exclusively concentrated in New Zealand (66% of volume, 37% of value) and Australia (28% of volume, 59% of value) and, unlike in other regions, it is mollusc cultivation (principally mussels in

New Zealand and oysters in Australia) rather than diadromous fish farming that dominates.

Fish are an important protein source for many Pacific islanders—supplying as much as 77% of the Solomon Islands' and 85% of the Maldives' animal protein needs—although culturally and nutritionally it is of lesser importance in both New Zealand and Australia. While under-reporting tends to bias supply figures downwards for many of the Oceanic countries, nevertheless 2004 per caput supply ascended to 185.9 kg in the Maldives (the highest in the world), 75.2 kg in Kiribati and 94.3 kg in Palau.

REASONS FOR PRODUCTION GROWTH

However, the CAGR for other leading fish-producing coastal states has been 4.85 per cent for West Bengal, 4.47 per cent for Tamil Nadu, 3.37 per cent for Gujarat and 3.25 per cent for Odisha between 2004-05 and 2008-09, it said. Interestingly, non-coastal state of Haryana had registered the highest growth rate of 14.7 per cent in fish production followed by Chattishgarh (11.29 per cent) and Uttar Pradesh (7.05 per cent), the industry body said. The factors that have contributed to the development of world capture fisheries over the last half century, a period that witnessed a rapid growth (circa 6% p.a.) in landings during the 1950s and 1960s, but a much slower rate of increase in the subsequent three decades. Indeed, as we saw earlier, in some sea areas there has been a leveling off or decline in catches for certain groups of species (e.g. cod from the NE Atlantic), mainly due to stock depletion caused by excessive pressure of fishing. In our discussion, therefore, we will consider how various factors (some positive, some negative) have influenced global fisheries production.

The Technological Revolution

An important supply-side driver has been the developments in fishing technology. While the switch from sail to steam power and the introduction of hydraulic landing gears permitted the deployment of large otter trawls and seines prior to the First World War, the pace of technological development in the sector has accelerated over the last half century. Echosounders and netsondes were developed for use on vessels and fishing gear in the late 1940s and 1950s, enabling fishers to locate—and then position fishing gear with more precision over—schools of fish. More recent acoustic innovations, such as the multi-frequency towed device developed by CSIRO have, by allowing scientists to distinguish between both species and sizes of fish, permitted the more accurate assessment of the biomass of species such as the orange roughy and blue grenadier, and it is probably only a matter of time before such devices are deployed across the industry so as to further enhance catches.

The introduction of synthetic fibres in the 1950s not only permitted the development and rapid dissemination of power block seining, but made mid-water trawling and gill-netting (a common fishing technique across much of the developing world) more effective, while underpinning the emergence of modern long-lining techniques. Fleet improvements have also contributed to an increase in fishing power, with relatively high-cost steam-driven trawlers being crowded out of many fishing grounds by diesel-engined trawlers with cheap hydraulic winch systems by the 1970s. Improvements in hull shape and vessel design have proceeded apace. The deployment of navigational aids such as Global Positioning Systems (GPS) has allowed the location of fishing areas or topographic features to be pinpointed with greater accuracy and the whereabouts of vessels to be determined within tens of metres. GPS, in conjunction with digital charts and sidescan sonar, enables more detailed seabed mapping and the construction of highly detailed three-dimensional models of fishing grounds. Incorporating information on time and tides, weather conditions and past catches further improves the effectiveness of fishing operations. Post-harvest developments, in particular improved freezing and at-sea processing technology (case of industrial fisheries) and the more widespread distribution of ice-blocks (artisanal fisheries) have also enabled fishers to extend the range and duration of fishing trips.

Integral to these developments has been scientific research. Growing knowledge about fish behaviour has been complemented by ever more precise measurement and calibration of gear performance. In the 1960s commercial gear development research led to the development of more efficient trawling techniques and fish/shellfish selection. The impact of fishing gears on the sea-bed, twine thickness and improving the energy efficiency of fleets in the wake of the 1973 oil crisis came to the fore in the 1970s, with the development of species and size-selective commercial fishing techniques being given a high research priority in the following decades. These technological developments have facilitated both the intensification and extensification of fishing, increasing the efficiency of fishing and expanding the areas that may be fished. Presently, for example, modern long-lining gear generally comprises a 60-80 mile main-line and up to 3000 baited hooks suspended off a series of branch-lines, 40 mile drift nets are commonplace, while the maw of the Gloria supertrawl net is large enough to accommodate twelve Boeing 747s.

While technological change has been a dominant force in the expansion of world capture fisheries, it has also almost certainly contributed to the overexploitation of many fish stocks and in some cases led to their near collapse. The likelihood of this happening has been greater in fisheries where the regulatory controls have been weak or ineffective, and where increases in the fishing power of vessels have been allowed to take place without regard to the consequences this might have for the effective catching capacity of the fleet. The post-war development of the NE Atlantic herring fisheries provides a salutary illustration of this, with the rapid adoption of new fishing methods (notably pelagic trawling and purse seining) during the 1960s resulting in rapidly rising catches but, eventually, to stock depletion and the closure of the fisheries in the North Sea and West of Scotland. That said, not all technological developments have been to the universal detriment of fish stocks. The introduction of vessel monitoring systems (VMS) has aided regulatory authorities to ensure skippers comply with national/international fisheries regulations , although their use has not yet gained worldwide acceptance. Equally, the growing sophistication of fishing gears has helped reduce by-catch and the ensnarement of non-target species.

Government and Donor Support

While government and donor support to the sector has historically taken a variety of forms, such support can fall into one of three categories. Direct payments such as price support payments, grants towards the purchase of new fishing vessels and/or gear, compensation payments for disrupted fishing seasons, and disaster relief payments etc. These enhance the revenues of the recipients and are paid from government or donor budgets directly to the fishers. One example of such support was the September 2006 Irish government announcement of almost €56 million of government and EU grant aid for the introduction of 38 new vessels into the Irish Whitefish fleet, including a special €3 million scheme designed to assist young fishermen to obtain their first vessel.

Cost reducing transfers such as fuel tax exemptions, subsidized (or free) ice, gear and vessel provision, preferential rates of interest for vessel/gear purchase, and reduced charges by government agencies etc.

These are paid by the government or donors to fishers to reduce capital and/or running costs. Over-zealous aid agencies, for example, have contributed to the pressure on fish stocks in the eastern Indian ocean by supplying too many new boats to coastal communities hit by the 2004 tsunami. While Thai geo-informatic agency estimates indicated that 4700 fishing boats were destroyed across 406 Thai fishing communities, Thai Fisheries department calculations suggested a total of 24 400 boats have been replaced (the majority of which had been donated) or repaired!

General services refers to all other support schemes, such as the construction of port facilities, research expenditure, restocking costs, market intervention schemes, protection of marine areas, regional development grants, managerial and technical support to fisher organizations etc. These benefit the industry, but the costs are not fully recouped from the sector. One illustration of this was the modernization of the Cotonou fishing port in Benin in 2004.

Costing US$7.3 million and benefiting some 4000 fishermen, the costs of the scheme were largely met by the Japanese International Cooperation Agency (92.38%) rather than the fishermen themselves. While estimates as to the annual level of support vary widely—the OECD mentions a level of around US$6 billion across OECD member states, and the World Bank suggests a global figure of between US$14 and 20 billion—what is undeniable is that much of this support, by reducing the private costs of fishing activities, has encouraged fishers to trade up vessels and/or gear and expand fishing activities with a consequent deleterious impact on stock levels. A belated recognition of this, allied to the precarious nature of many of the major commercial fish stocks, now sees support being (partially) reoriented towards the decommissioning of vessels, the buy-out of licences and permits, the introduction of re-training schemes and a raft of other measures designed to reduce the pressure on fish stocks.

The Attraction of Fishing

Fishing has proved a relatively attractive commercial and/or subsistence occupation for new entrants in a number of instances, notwithstanding the widely-held thesis that fishers are among 'the poorest of the poor'. Reasons for this are manifold. In Panama, for example, the mechanization of sugar and rice production led to a massive influx of displaced agricultural wage-labourers into the artisanal shrimp-harvesting sector between the 1960s and mid-1980s. The Chilean devaluation of the early 1970s attracted substantial inward investment into the industrial fishing sector, with one beneficiary claiming to have recouped their US$200 000 investment within the space of a week. A reversal of the anti-export bias in Argentine macro-economic policy, allied to tax exemptions on new fishing vessel purchases and a relaxation of the rules governing entry into the industrial fishery, facilitated a dramatic growth in the Argentine offshore fleet in the 1990s. In Canada, the availability of insurance under the Special Seasonable Fisherman's Benefit Programme accounted for the presence of one-half of the inshore fishing boats in Newfoundland, and postponed a restructuring of the industry until the 1990s.

All the above examples do share one thing in common, however. In each instance, fishing—given the prevailing technology—is more attractive than alternative employment and/or investment opportunities, and so entry into (or exit from) the industry is encouraged (discouraged). While the relative attractiveness of the industry can be modified by shifts in government/donor policies and access regimes (and in the longer term by technical change), linkages between the sector and other labour markets and investment opportunities (at the local, national, regional and global levels) can also have a material effect upon fishing effort and consequently, stock sustainability.

The Property Rights Regime

A third supply-side factor encouraging the over-exploitation of fisheries resources is the nature of the underlying property regime governing access to fish stocks. While land is relatively easy to demarcate, territorial rights at (and over) the sea are less easy to define/enforce and so, by default, open access fisheries were historically the norm rather than the exception. This principle was formalized by Hugo Grotius in 1608 who coined the phrase *Mare Liberum* the 'Freedom of the Sea', which has dominated humanity's approach to fishery resources for the last 400 years. In open access fisheries, serial depletion of species is routine. The same tendency is also apparent under common pool property regimes—where resources pertain to a defined community—members seeking to maximize their gain by harvesting more (and more) fish. The 'race for fish' that develops encourages over-investment and results in the progressive over-fishing of fish stocks—with species only escaping over-exploitation if their range extends beyond the scope of current fishing technologies—an increasingly finite option, and/or the open access regime is terminated. Since the mid 1980s, increasing attention has been orientated to property rights regimes in fisheries, and the idea that the assignation of private property rights can resolve the problem of over-fishing. Current research, however, suggests that the allocation of private property rights via quotas may not necessarily resolve overfishing problems, but could in fact foster the concentration of capital and an increase in by-catch.

Changing Consumer Tastes

An increased attention to diet and personal health over the last half century or so has also stimulated consumption demand for fish. Fish is low in saturated fats, is a rich source of high-quality protein and Omega-3 fatty acids, and also provides essential minerals (iron, calcium, zinc—from shellfish—iodine and selenium) and vitamins (B and D) which are easily absorbed by the body. Omega-3 fatty acids are now known to be particularly beneficial in terms of cardiovascular health, for example, and some studies have indicated that early exposure to Omega-3 fats may also enhance brain and retina development. The Iowa Department of Public Health in USA, for example, suggests that eating fish may protect against a variety of diseases and illnesses in adults, including cancer, dementia, diabetes, depression, rheumatoid arthritis, psoriasis, prostrate cancer and autoimmune disease.

While information and advice like this has undoubtedly stimulated the demand for fish products—a demand which was aided by the relative historic cheapness of the product compared to alternative protein/food sources such as red meats—it has been tempered in recent years by public health concerns regarding the levels of environmental contaminants encountered in fish/ shellfish. Some farmed salmon and fatty fish have been found to have significantly high concentrations of PCBs (polychlorinated biphenyls) and dioxins, while high levels of methyl mercury have been encountered in predatory marine (tuna, swordfish, shark and halibut) and large freshwater fish. As a consequence, many food standards agencies in the developed world have issued advice on what they consider to be 'safe' fish intakes. In the developing world, consumption of raw or inadequately cooked fish can also pose serious threats to health, with medical complaints such as bile duct cancer, gallstones, diarrhoea, and peptic ulcers increasingly being attributed to fish-borne zoonotic parasites in Southeast Asia.

Although tastes have not changed in China and Southern Asia, where fish has always been a consumption item of choice, rising incomes in the region have nevertheless also contributed to growing consumer demand. Currently, global fish consumption stands at around 15 kg per capita—double the level of thirty years ago—and has been instrumental in underpinning the growth and internationalization of fish commodity chains. The international fish trade grew nine-fold in the twenty years from 1981-2001 (from US$6.1 billion to US$56 billion), for example, with a particularly sharp rise in the receipts accruing to developing countries (up from US$3.4 billion to US$17.4 billion) over the same period. Moreover, according to the same author, this growth in international trade has not had a detrimental effect on food security in the form of fish for food—although the same cannot perhaps be said for fish stocks.

ESTIMATION OF MARINE FISH PRODUCTION

The marine fish landings in India during 2002-2003 was estimated at 2.64 million tonnes, which is 0.314 million tonnes (13.5%) higher than that of the previous year. The increase was primarily due to enhanced catches of sharks, oil sardines, Bombay duck, ribbonfishes, carangids, seer fishes, tunas, penaeid prawns and cephalopods. Perches and non-penaeid prawns showed a decrease in landings. The mechanized sector accounted for 67.9%, motorized sector 25% and artisanal sector 7.1% of the production. The north- west coast accounted for 0.908 million tonnes, followed by south-west coast 0.86 million tonnes, southeast 0.611 million tonnes and north-east 0.227 million tonnes. Monitoring of environmental characteristics of coastal water was continued all through the year. Plants form the base of the food chain and therefore play a major role in controlling the overall productivity and function of freshwaters. They also provide structure to the lake or pond and create a diversity of habitats that are essential to many animals including fish. It is important for those managing fisheries and freshwater ecosystems generally to understand the environmental factors that control the distribution and abundance of plants. It can be simply stated that, generally, the more plants in your fishery the greater its ability to support other animals and the fish living within it.

Plants help create the natural balance that should be the aim of all fishery managers. Fish that live in environmental conditions that are as near optimal for them as possible will be less stressed than fish that live in conditions that are suboptimal, and plants are the first and most important step in creating this balance.

Plants have several important functions in the ecology of a still-water; these include:

- the provision of habitat;
- serving as a substrate for spawning;
- acting as a food source;
- water filtration and purification;
- the production of oxygen by photosynthesis;
- regulation of the nitrogen cycle.

Plants also have a function in protecting banks against erosion caused by wind and wave action as well as human and animal impact such as that produced by simply walking along a bank. Marginal and emergent aquatic plants form the all-important intermediate zone or interface between the aquatic and terrestrial environments. It is important to appreciate that some species of plant can become a nuisance owing to their invasive nature or to other undesirable impacts such as those caused by algal blooms. Before considering what plants are beneficial and what purpose they serve, it is essential for the fishery manager to have an understanding of the types of aquatic and semi-aquatic plant.

Two main groups of plants are significant in freshwaters: algae and macrophytes. The two groups form separate but linked components of these ecosystems as they have very different requirements to flourish and equally have very different functions and impacts within a system. The capacity to create their own food through photosynthesis is the single most important characteristic of plants and distinguishes them from other forms of life.

ALGAE

Algae are unicellular or multicellular rootless plants that never differentiate into stems or leaves. They grow in proportion to the amount of available nutrients, oxygen concentrations and sunlight. They can affect water quality by both increasing and decreasing the amount of oxygen dissolved in the water. They are an essential form of food for fish and small aquatic animals. Algae are found in most habitats on earth, though most occur in freshwater or marine environments. There are thought to be around 5000 species of algae that inhabit freshwaters in the UK.

Morphology

Algae come in a variety of shapes and in varied colours due to their different photosynthetic pigments. Algae can be unicellular and microscopic or colonial, forming plate-like colonies, thread-like filaments, net-like tubes or hollow balls. Many planktonic algae species bear horns, ridges or wings to increase their surface area to volume ratio, which not only increases their ability to obtain scarce nutrients from the environment, but also protects them from herbivores and slows the rate with which they sink in water.

Sizes of algae vary hugely. Diatoms may range in size from 2 ì m to several millimetres, although there are only a few species larger than 200 ì m; whereas colonial brown algae, in which some species form branched filaments or foliose plants, may be many metres long. Some algae, like diatoms, are encased in a siliceous cell wall, which takes the form of a box, and come in many unique and beautiful forms.

Reproduction

Algae reproduce either sexually or asexually. They use various methods that are fairly group specific, but some features are shared. When reproducing asexually, most groups develop spores within a parent cell, which are released into the environment. Each of these spores then develops into a single algal cell, or they divide by mitosis into numerous cells. For example, reproduction in the cryptomonads is asexual and primarily by longitudinal cell division with the cell dividing in either a free-swimming or non-motile form. Sexual reproduction is not rare but occurs less often than asexual reproduction. Gametes are produced in different algal cells and released into the environment. Male gametes usually attach themselves to female gametes and their genetic material is absorbed through the cell coat of the female. In freshwater species, this is followed by the formation of a cyst in which meiosis takes place. Cyanobacteria only reproduce asexually, with genetic recombination accomplished through transformation or conjugation. During transformation, DNA, released from donor cells, is incorporated into recipient cells. Conjugation is the process by which two cells become connected by a narrow tube, through which DNA moves from one cell into the other.

Algal groups and abundance

Algae usually make up a significant amount of the plant biomass in still waters. Larger algae, such as the filamentous algae that is often referred to as blanket weed, can grow densely in shallow water and provide distinctive habitats for other organisms. Microscopic algae are present as floating forms (phytoplankton) and sessile forms (epiphyton) attached to plant stems, rocks and even bottom substrate, providing the base for food chains that are particularly important for fish fry and zooplankton (small microscopic aquatic animals) in still-waters.

When present in high densities, phytoplankton may form an algal bloom. Blooms often result from a population explosion of phytoplankton in response to changing environmental conditions, including nutrient enrichment (eutrophication). Blooms can result in oxygen depletion and biological impacts. These blooms are often visible as a discoloration of the water; the degree of water coloration can, with experience, indicate the density of the algae in the water column. Given that most still-waters are eutrophic, it is easy to understand why it is difficult to manage still-waters that have a predisposition to algal blooms.

MACROPHYTES

Macrophyte form and function

Macrophytes are the group of plants that most fishery managers will be familiar with, as they are generally much larger than algae, have a defined structure and can be easily seen in some shape or form along the banks of most ponds, lakes, rivers and canals. The types of aquatic plant are extremely diverse, but they can be separated into four categories based on their habitat. These are riparian, emergent/marginal, floating (attached and unattached) and submersed.

Higher plants evolved from primitive algae or algae-like ancestors, but probably on land, not in the water. Today ' s aquatic macrophytes are specialised forms that have recolonised aquatic environments. Because aquatic angiosperms (flowering, sexually reproducing plants) are closely related to terrestrial types, it is not surprising that the morphology and anatomy of these two groups are basically the same. However, their differences become obvious when considering the adaptations that aquatic species have evolved to survive in their watery environments.

The flowers of most aquatic angiosperms must be elevated above the water for pollination to occur: entomophily is pollination by insects and anemophily is by wind; only very rarely does pollen transfer occur underwater (hydrophily). Getting and keeping flowers at the surface requires special adaptations; water lilies, for example, have waxy, bowl-shaped flowers that float by themselves, much like a small boat; some submerged plants such as water milfoil (*Myriophyllum spicatum*) develop rigid stems that protrude above the water and on which small flowers are borne. Aquatic macrophytes tend to replace sexual reproduction by vegetative reproduction, which may be related to the difficulty in raising the flowers above the water for aerial fertilisation. Vegetative, or asexual, reproduction is a vital key to survival among the aquatic plants. Some species rarely generate viable seeds and those that are produced serve more as a ' back-up ' to ensure the species ' survival in the event of a disaster. Vegetative reproduction occurs primarily by stem fragmentation, which is one of the primary methods used by Canadian pondweed (*Elodea canadensis*), but some species such as duck weed (*Lemna* sp.) use the whole plant, hornwort (*Ceratophyllum demersum*) uses its shoot fragments, and specialised organs such as tubers are used by yet other species such as *Potamogeton* sp.

Macrophyte types Bankside (riparian) plants

Many plants occupy the dry land of the bank, although their roots may be in the ground-water surrounding the lake or pond. Most conspicuous of the bankside plants are trees, particularly ones resistant to flooding such as willow, alder and poplar. Around small ponds trees may occupy valuable space and their leaf litter may cause water quality and siltation problems. On larger lakes they are a feature of the landscape that provide shade for fish and are aesthetically pleasing for the visitor. Various forms of grass, low shrub and wild flower are also common features on banks and may help stabilise the edges of the pond or lake.

Emergent and marginal plants

Emergent plants are those whose roots grow underwater, but their stems and leaves are found above the water. Emergent plants are also sometimes described as ' marginals ' because they grow in the shallow waters along the bank. They provide useful habitats for many fish species and for a whole variety of invertebrate life such as dragonfly nymphs, water snails, etc., and as such are to be encouraged. A further beneficial factor is that the emergent plant stems confine water movement and aid the warming of the margins. This makes this marginal area very suitable for the establishment of dense zooplankton communities on which juvenile fish can feed. However, in a recreational fishery, they may need to be strictly controlled as they may make access to the water difficult and can snare fishing gear.

Floating plants

These are plants in which part or nearly all of the plant floats on the surface of the water. Floating unattached plants, such as duckweed, are those in which most of the plant is at or near the surface of the water. Roots, if present, hang free in the water and they are not anchored to the bottom. Floating attached plants have leaves that float on the surface, but their stems are beneath the surface, and their roots anchor the plant in the substrate. Some floating plants can be a nuisance as they can expand rapidly over the surface of the water body, shading out the water column so plankton cannot grow and limiting access by fisher-men. Floating leaved plants such as water lilies form valuable shade and resting places for fish.

Submersed plants

Submersed plants are found when the entire plant is below the surface of the water. Their growth is affected by water clarity and they can often be the subject of overgrazing by the fish community, especially if this includes a population of carp. Submersed plants are particularly valuable as spawning sites for many species of fish and provide refuge for the fry once they hatch.

DEVELOPING A GREEN-FIELD SITE

Often developing a fishery from scratch will require the development of a green-field site. However, before any development takes place a suitable site must be found. In many respects this can be the most difficult aspect of the entire procedure of developing and creating a fishery.

A fishery needs land and adequate good-quality water. The UK is a very densely populated island. In the past, its freshwater resources have been adequate for supporting the population with drinking water, and water for agriculture, industry and recreation. However, even in the past, stresses on the resources were such as to result in severe pollution and local shortages in supply. Now, with changing climate and growing population pressures, the quality and quantity of water is constantly under pressure in parts of the UK as rainfall appears to be declining and becoming less predictable. Whatever the cause and the degree of impact that this will have on the environment, it looks certain that the current trend towards heightened water awareness by the general public is set to continue for the foreseeable future. It is this awareness that will drive those organisations responsible for water regulation to be increasingly vigilant when considering applications for abstraction and discharge consent. The need for continuity and protection for the potable supply carries immense political weight.

Obtaining a plot of land suitable for development is not all doom and gloom, far from it. As this is still a wet island, standing waters occur frequently in lowland areas. It is, however, probably fair to say that any future developments will need to have an increasing regard for their potential impact upon the environment.

The bureaucracy involved in turning a green-field site into a fishery is considerable; however, before considering this it may be pertinent to consider some aspects of locating a suitable parcel of land for development.

Finding and identifying a green-field site suitable for a fishery

Before identifying a suitable site and obtaining some form of tenure (licence, lease or purchase) on it, one should consider exactly what is needed, both in terms of what is essential and what is desirable in the site. Ultimately the type of fishery that is to be developed will be the main criterion for deciding the suitability of the site, although there may be some flexibility to modify the type of fishery to suit the site. For example, certain water-quality criteria are vital in determining whether a fishery would be suitable for trout. Certainly water quality is not a factor that should be compromised: generally if it is not suitable to begin with, in practice it will be difficult to improve at a later stage.

Finding the right site in a convenient location is very important. Only investigate sites within a predefined radius of your base. Consideration should be given about what is a reasonable distance to travel on a daily basis. Of course, this decision is a matter of personal choice, but it stands to reason that, as most people live in the urban environment, then most will have to travel to rural areas to find and locate a green-field site. There is little point in coming to an agreement to develop a site many miles from home only to find that the strain of travelling is too great to sustain. The actual finding of a potential site requires a great deal of personal effort. It is a very time-consuming procedure. Although some fishery owners will surely report that they just ' came across ' their site, most sites are found after a great deal of time and effort. Time and effort are required because there is nowhere that one can go to find out where a defined green-field site awaiting development is situated. There are many ways of obtaining information that will make the task of finding such a site easier. These may include speaking to local people; success may often depend on word of mouth and who you talk too. However, local knowledge should not be overlooked.

There can be no doubt that the best method in condensing the search area is the use of Ordnance Survey maps and the ability to read and understand contour lines. If this is coupled with a map showing geological and hydrological features, much of the initial detective work can be done from the comfort of your own home. The Ordnance Survey's 'pathfinder' 1:25,000 series of maps is well suited. They are available from bookshops within the local area to be investigated. The area should be near to a water supply from either running or spring/groundwater sources; it should have an impermeable substrate and be accessible. If the area broadly fits these criteria then it will merit further investigation, and it is time for a site visit. On-line searches using geographical engines such as Google Earth or Google Map can also help refine the search and clarify the setting around an identified area. Many of the classic signs of a suitable site will only ever be found by close observation in areas of potential site development. These may include the following:

- standing surface water;
- boggy appearance;
- *Juncus* (soft rush) and *Carex* (pond sedge)-type grass tufts or other aquatic plants;
- nearby water source.

SITE ASSESSMENT OF A SUITABLE GREEN-FIELD SITE FOR A FISHERY

Once a potential site has been located, a decision needs to be taken about the site's suitability for the proposed use. To decide this some further information needs to be drawn together for consideration. It is suggested that the following factors be considered when assessing a green-field site for its suitability for a particular type of fishery:

Access

The site will not be viable if access makes either the construction or the future management of the facility difficult and expensive to undertake. It is advisable to take a very detailed look at this aspect early in the site visit. The laying and maintenance of a hard surface suitable for vehicular access can often be a very limiting factor both logistically and financially.

Site security

The potential security of the site is easily overlooked. There can be little point in successfully creating a fishery only to see the fruits of one ' s labours disappearing by theft (poaching) and predation (herons, cormorants, mink, etc.). An isolated location will make the security of the site very difficult. The theft of fish is widespread, even from angling waters where fisher-men are in attendance. A near-ideal site might of course have accommodation on site or very nearby. Every site is unique. It is important to consider site and stock security as one of the contributing and deciding factors in a decision about suitability for the purpose.

Site availability

There is little point in formulating plans to develop and construct a fishery if the potential land is not commercially available. This needs to be confirmed at an early stage. Local knowledge can be an advantage: a question at a local pub, or to a passer-by, as to the ownership of the potential plot of land may be helpful. It is advisable to keep the amount of information departed to a minimum: enquire the name of the owner without giving your reason for wanting the information, other than to contact them. It is not unusual to find a piece of land that the local inhabitants cannot identify an owner for, as often with suitable sites they are poor agricultural land that is left fallow for year on year. The local farmer may well be your next best option. Do your research first as they might turn out to be the owner. An additional source of ownership information is through the Land Registry.

Nature conservation

Increasingly, many parts of the countryside are being protected by various conservation -status designations. It is important to establish as early as possible if any of these designations affect the potential site, or indeed if any are up-or downstream. The designation can affect whether development would be allowed and, if it were, whether the management regime would be required to coexist with the conservation status.

The development of an area that has been designated with a special status will as a minimum undoubtedly involve outside bodies such as national and local conservation bodies. If the area is designated as a Site of Special Scientific Interest (SSSI), then Natural England will need to be consulted.

Soil types

Fortunately, the tell-tale signs that may lead to a site investigation in the first instance can also be indicators of the suitability of the soil for holding water, namely visibly wet and swampy ground. Fisheries rely on a natural bottom and sides to the pond or lake, therefore the material that is left after construction must be impermeable to keep seepage to a minimum.

The suitability of soil for a fishery is more involved than just its ability to retain water. The soil richness will directly reflect in the fertility of the pond and thus its ability to support a balanced aquatic ecology, including a fish community. The best way to establish suitability is to perform a soil survey.

Soil surveys should be done as early as possible. Soil sampling can be carried out in two ways, which are dependent on the analysis required from the sample. They are:

- disturbed samples;
- undisturbed samples.

Disturbed samples do not represent exactly how the soil lies in its natural state (before sampling), whereas an undisturbed sample represents exactly how the soil lies in its natural state.

The disturbed sample is used for the simple tests: those that can be performed at the site while making the initial site visit. The disturbed-sample method also includes a simple test, which can be done at the site to establish soil texture and its ability to retain water.

It follows that the undisturbed sample is used for the more sophisticated tests, which are more suited to laboratory investigation, such as chemical analysis. The undisturbed sample requires a more involved procedure to obtain it. Before developing a green-field site for a fishery, it is suggested that both types of sample will be required and the appropriate analyses performed.

The initial survey may be termed the ' reconnaissance survey ' . It should be quick and provide a general idea of the soil varieties present and where they are found. A survey like this should be conducted by digging several open pits and then examining the exposed soil profile. From this it should be possible to determine what parts of the site are suitable for pond construction (those having good impermeable soil), and those parts that are not suitable (gravel outcrops, thick layers of organic material, etc.).

Close inspection of the open pit will reveal the soil profile. This comprises the soil horizon (layers of different types). It will be worthwhile making a sketch or taking a photograph, and taking a sample of each layer. This is not quite the same as an undisturbed sample, as the soil is only lying in its natural state in the horizontal plane.

It is worth noting here that should soil be required to construct a dam or dyke, it needs to have the cohesive properties or plasticity index of 15. Soils that contain too much sand or gravel should not be used because they will not hold water. Clays and silty clays are best. Soils that contain at least 20% clay by volume are suitable for making impounding banks.

It is clear that a soil ' s ability to retain water and to act as a nutrient base is important to the creation of ponds and lakes for recreational fisheries.

WATER SUPPLY AND QUANTITY QUANTITY

Although we call the ponds and lakes we use for our fisheries ' still-waters ' , the water in them changes constantly. The amount of time a mass of water remains in the still-water is called the ' retention time ' . This can vary according to the nature of the water body. For example, a lake created by damming a river will have a short retention time compared with an isolated pond on a clay soil. The amount of water in any still-water is the result of a balance between water coming into the water body (gains) and water leaving the water body (losses). It can also vary seasonally, with much shorter retention times in the rainy autumn and winter seasons than in the summer. On first impression, water quantity may appear to be a straightforward criterion. Indeed, if the supply is consistent and obviously of suitable volume, this may be true. Where the supply is more limited it may be necessary to estimate how much water will be required. The amount of water that will be required to fill a fishery should be worked out in cubic metres (m^3).

Gains

- Inflow from rivers or streams: usually the main source of water. If water is to be deviated or abstracted from a nearby river, an abstraction licence will be required unless there is already a permit in place.
- Seepage from ground water: especially in low-lying swampy areas, or ponds located in valley bottoms. The rates of seepage in or out of the pond depend on the nature of the soil and bedrock in which it is located.
- Pumping from ground waters: a locally important source of water although abstraction licences will have to be applied for.
- Local rainfall and surface flow in the lake catchment: usually a minor condition but can be seasonally important in areas liable to flash flooding.

For small ponds with a generous water supply (an annual flow more than six times the pond's volume) that flows all year, volume alone (in cubic metres) will be sufficient to determine the adequacy of water supply. For larger ponds with a less generous water supply (an annual flow less than three times the pond's volume), it may be prudent to calculate the seepage and evaporation factors as well.

Losses

- Outflow into rivers or streams: this can be controlled by a series of flow control structures.
- Seepage into groundwater.
- Evaporation: losses by evaporation are variable and depend on air temperature and wind.
- Abstractions: usually not important in the types of still-water that are of interest.
- Evaporation in a dry summer month can account for a 200-mm loss in a 0.4-hectare pond (equivalent to about 7 mm/day). In a fishery, it is the annual flow that is in many ways the limiting factor. In a still-water, a guide might be that a minimum of twice the water volume required should be available as the annual flow.

The estimation of the quantity of water available is generally dependent on flow. It is important that a fairly accurate estimation of water flow can be gained under field conditions. There are several ways to measure the amount of water in a stream or channel. The method that you choose may depend on several factors:

- the accuracy of the result needed;

- the quantity of water present in the stream or channel to be measured;
- the site access;
- the equipment available.

Under normal circumstances an initial site investigation should be done before approaching the landowner. It follows that, before making such an approach, one would wish to have a very good idea about the continuity and volume of the flow. The float and cross-section method will give the required degree of accuracy and is fairly simple and quick to perform with the minimum of equipment.

Type of supply

An initial investigation should quickly reveal the source of the water supplying the potential still-water. The three likely sources are:

All provide very different types of water quality. Perhaps the most important point to note at this stage is that a groundwater supply will require either drilling a borehole or excavating the pond or lake down into the water table. Excavating into the water table can only ever be a viable option where the level of the water table remains stable during the summer months.

To develop a borehole, initial test drilling is needed before a decision is made for one to be sunk permanently. This can be a very expensive operation, requiring specialised equipment and operatives. It will, if settled upon as a method of supply, also require pumping, and the associated costs will have to be met as part of any running cost. So water quality potentially may well have a trade-off against additional cost.

Before any drilling is undertaken, the Environment Agency must be approached as there are many areas where the sinking of a borehole will not be allowed. The way forward is to have an informal discussion with the water resource function of the Environment Agency. Establishing a borehole may be considered as having an impact on the aquifer. This is particularly true in the southeast England, where a large percentage of the potable supply in some areas comes from an aquifer, which may struggle to meet the demands placed upon it.

Water quality

Having confirmed that the soil can retain the water, and that there is enough water, we must ascertain that the quality of the water is suitable to support fish in a still-water fishery. The quality of water can be simply tested and involves taking a sample in a clean, dry plastic bottle. The sample can be tested with a commercially available test kit. Some parameters such as dissolved oxygen, pH and temperature can very easily be taken with small electronic probes. Once you have analysed the water sample, you then need to be able to understand what the analysis means in terms of creating and sustaining your fishery. A single sample test is only a snapshot in time, indicating water quality at the time of sampling. For this reason further evidence about the viability of the supply, particularly upstream pollution, will prove extremely useful. In many instances there may be some previous history of poor water quality, so the fishery manager, having observed the site over many months, will be able to determine that there is an adequate water supply.

In the absence of previous knowledge it is possible to gauge frequency and severity of pollution events by sampling the animals living there (biotic taxa) as an environmental indicator.

Larger carriers (rivers and streams) may have had similar sampling done by the Environment Agency. It is always worth a call to your local Environment Agency fisheries officer or area biological appraisal teams to ask if any information is available. Also, remember that any upstream fishery can impose a potential disease problems. The Environment Agency may also have record of any known risks in the catchment.

The topography of the potential site is the factor that dictates the ease with which a pond or lake can be constructed. Ideally it should suit low-cost construction and should not be subject to risk of flooding. An ideal site will also facilitate the emptying of the fishery without pumping (that is, by gravity). The objective must be to exploit the particular advantages of a given site, while ameliorating the disadvantages, in order to maximise the most cost -effective construction method.

OBTAINING THE SITE

Having decided that the site has potential, it is advisable to define the precise objectives of the scheme. This will enable a plan to be drawn together, which should be used as a guide once the site has been obtained. The plan can also form the basis of a management plan that could be presented to a financial institution to raise any funds that may be required. It will be almost impossible to progress past this stage on a commercial basis without a set of comprehensive plans and accompanying cash flow and budgets.

A set of ground plans showing the site layout with access, etc. will be required by the planning authorities. The Environment Agency will, if appropriate, require fairly concise details about abstraction and flow through (discharge) as well as details of any impoundment and associated structures such as dams. However, before spending money on these aspects, the purchase or lease of the site must be secured first.

The owner will have to be approached. To a large extent, how the approach is made will depend upon the person making the approach and those being approached. A personal approach can often be beneficial but many may be ill at ease in doing this, so an initial well-written letter may be better. Obviously, a lot depends on how it is envisaged that the recipient will react.

The buyer should be armed with a strategy, and know all of their options including what they are prepared to pay for the site; then, and only then, should the approach be made. It may also be advisable to have put the finances in place or at least know where they are coming from.

Options for securing the site

You must also know your preferred option for securing the site. These include the following:

- a shake of the hand;
- a lease;
- a licence;
- purchase.

A shake of the hand

A shake of the hand or gentlemen ' s agreement is a quick and easy way to reach an agreement to establish a right to construct your fishery on the proposed site. There is, however, one obvious major disadvantage. There is no security of tenure. Your being on the site is dependent solely on an agreement made orally. Furthermore, the agreement is generally made with a specific person and this may be negated should the ownership change.

Any time and effort involved in a ' shake of the hand ' agreement should be minimal compared with the return. This does not sound like a situation very likely to apply to the development of a fishery.

Leasehold

A leasehold will give a predefined period of security. A renewal clause could (perhaps) be written in that allows the tenant the right of renewal at the end of that period. The period, wording, renewal, etc. of a lease is a matter of negotiation.

It is advisable to consult a solicitor. Remember that, although a lease may give the tenant security, it will be in the mind of the landlord ' s solicitor to ensure security, protection and indemnity for his client, the landowner. It should also be remembered that the person or persons signing the lease are responsible for all the terms contained therein for the period of the lease, and sometimes on surrender. This will be regardless of any circumstances that may affect the operation the fishery.

It is given these harsh realities that many potential fishery managers will decide for various reasons that the risk outweighs either their confidence/ability, or the potential gain, or a combination of these and other factors.

It should be noted that before giving up it is always worth trying to negotiate a lease with a ' get-out ' clause in the areas that are of concern. An example of this would be if the inflow of water was only about the minimum required and no history of seasonal and annual flows was available; then it may be possible to sign a long lease, with an option or ' get-out ' clause that covered the volume of water. Of course, it may well be with such a marginal site that it would be questionable if the development cost would not be outweighed by the risk.

Licence

Licences have grown in popularity in recent years. The legal definition of the difference between a lease and a licence is often to do with security and length of tenure. A licence is generally for a shorter term. It confers a permission to perform the activities detailed in the document for the defined period of time. It does not carry the legal weight of a lease; it is not as easy to rely on the document to resolve disputes in law between the two parties.

It is believed that the popularity of a licence has occurred because it can be put into place more easily without involving solicitors. It may be that there is a connection here with the recent trend for landowners to employ land agents to deal with such matters as negotiating land deals. The land agent will produce a suitable licence without involving a solicitor. When solicitors are involved the favoured document appears to be a lease. The lease or tenancy is a lengthier document and is thus more expensive to produce and be advised upon by the legal profession. This is not necessarily a bad thing.

Purchase

The purchase of a potential site is the dream of many who would wish to own a fishery. In most cases, however, even if there is available capital there will not be sufficient money to complete the purchase, the development and the set-up costs.

It is assumed that most of those wishing to create a fishery in the UK will have to do so within conventional methods and that any financier would wish to see a realistic ' return on investment ' . If a presentation can be made to a bank or other source of finance that shows there to be sufficient margin to cover the purchase, set up, running costs (start-up to opening), then if the return on investment is sufficient compared with the risk and investors may be interested. Commercial mortgages are available for land purchase; they would generally be made in stage payments in-line with site development. This shortfall and lag in funding would need to be bridged (financed from another source).

When to sign on the dotted line

Although the agreement of the owner of the potential site for development has been obtained, as yet other permission to develop the fishery will not have been obtained. Signing of any binding agreement at this stage would be very risky. It may be useful, however, to have a solicitor draw up a letter of intent. A letter of intent will lay down an agreement of both parties that if certain criteria are met, then there is a contract. This should be a simple document that states that, if the necessary permissions are forthcoming, the interested party will have first option on a lease or purchase, etc. Avoid signing any binding agreement until permission has been granted to undertake the proposed activity by all the relevant authorities. This is stressed, as there can be little point in having local planning consent if you are refused abstraction and discharge consents. It is surprising how interest is suddenly awakened once the word ' s ' pond ' and ' fish ' are mentioned in some quarters. At this stage it is advisable to keep all your information very close to your chest.

Submitting and obtaining permissions

All the effort up to this stage will have been for nothing unless the relevant authorities and permissions can be gained. It should be pointed out that part of this ' seeking permission ' procedure could be done as soon as the potential site has been found and identified as suitable. However, although it is possible to put in formal planning permission for a site that is not owned, such action is unlikely to win the appreciation of the site owner. The procedures required are somewhat lengthy and involved; they are, however, within the scope of the layperson who has the patience to work through the forms with determination and attention to detail.

Plans must be formalised before approaching the regulatory bodies. At this stage a decision must be taken whether to involve a land surveyor; the surveyor would draw up the site plan. Depending on the size of the site, or perhaps more correctly the size of the proposed development, the cost can vary from a few hundred pounds to perhaps a thousand or more. Cost is also dependent on the firm or individual contracted. There is perhaps a better chance in negotiating a lower fee with an individual than a firm of land surveyors. Those prepared for a bout of hard-nosed negotiation may be pleasantly surprised at the discounted 'fixed rate' deals that may be achieved.

Regardless of who draws up a plan of the site, it must represent the vision of those who are to run and manage the fishery. Obviously this vision should consider expert opinion about water supply, siting and construction of the fishery and, importantly, the ongoing access to and around the site. Increasingly, plans for fisheries must consider the environmental impact that the development may have. If it is possible to create some useful habitats while developing the site, this should be shown on the plan. Local planning authorities now have conservation officers who will be made aware of planning consents; their opinions can carry significant weight. With the involvement of a local planning authority ' s conservation officer there may come, as the expression goes, ' out of the woodwork', a host of other interested bodies, which often include single-interest voluntary groups.

Before considering the various stages involved in obtaining the relevant permissions and consents, it is useful to compile a list of the information required for submission to the controlling authorities. This stage of the process can be very costly in terms of both time and money, particularly if information required by the specific authority is omitted from the original submission and a subsequent resubmission has to be made. Many of these authorities and regulators only meet at predefined times, and not to suit the applicant.

It is very useful to have informal meetings and conversations with the relevant authorities to gain a firm appreciation of their requirements.

It cannot be stressed too strongly that informal meetings and discussions should be used to obtain a feel for exactly what information is required and in what way its presentation may affect any outcome. Human nature will dictate that a planning committee may have far more trouble in rejecting an application that has been submitted in a manner in which the local planning authority itself has requested, and that has considered aspects that were highlighted by them during informal discussions. That is not to say that even an application fulfilling all requirements may not be rejected: there is often no consistency. It is suggested that in most cases where a total rejection is given, this will have been made known at an informal level.

It should be acknowledged that there will be areas and circumstances where no level of negotiation, flexibility or finance will make it possible to submit a plan that will get approval. From a financial viewpoint, the sooner this is realised and a swift withdrawal made the better.

FISH STOCK ASSESSMENT

A fishery manager needs to know what species are present in his fishery, how many of each species, the overall weight (biomass) of fish, how quickly they are growing and the general health status of the fish. The problem is how to obtain this information. It is difficult to see through the water into a normal fishery so the fish cannot be counted. It is almost always impractical to drain all the water out of the pond and count and weigh each individual fish. So we must, of necessity, sample the population to gain enough information to allow us to estimate the total number of fish, etc. This is a perfectly valid approach and can often give reasonably accurate answers, which can then be used to plan and evaluate the management of the fishery. Given that the estimation of fish stock levels in the still-water environment can be difficult, it is sometimes a good idea to use indirect methods of fish stock assessment. Indirect methods use records of things like the clarity of the water, the presence of aquatic plants and the status of invertebrate populations to give information on the fish stocks. For example good water quality, luxuriant growths of soft, submerged plant species and a diverse invertebrate population will indicate ' well-balanced ' fish stocks that are in equilibrium with their environment. This information gives a good impression of general health of the fish stock and the stock density relative to the carrying capacity of the pond. However, this type of information is not sufficient if you need to determine feeding rates or assess the need for supplementary stocking.

PLANNING

Before beginning any fish stock survey, very clear objectives must be set out and considerable thought must be given to how to achieve these. It is very easy to waste money and time through ill-conceived assessments that will not provide the information required.

The assessment process generally consists of two steps: (1) sampling the fish and (2) interpreting the results of the sampling to convert them into an estimate.

SAMPLING THE FISH

Capture techniques

Most fish stock assessments require that a sample of fish be caught, so obviously one of the most important decisions is how to go about catching the fish. This is not as straightforward as might be imagined. Each method has its benefits and problems and these need to be clearly understood before you start. Furthermore, each method will have specific problems associated with particular fisheries and again these need to be clearly understood before you embark on the survey.

Seine netting

One of the most traditional survey methods is seine netting, employing long-established equipment and techniques that have changed little in the general principles applied. The seine net is known to date back to ancient Greek and Egyptian times. Nowadays, natural fibres have been replaced with synthetic materials in the manufacture of nets, which has not only extended their useful life but had an impact on the efficiency of capture.

The seine net consists of a single sheet of meshed netting that is held vertically in the water by floats attached to the ' head rope ' or ' cork line ', which is fastened to its upper edge, and weights attached to the 'foot rope' or 'lead line'. When deployed, the net should form a wall between the surface of the water and the bed, encircling any fish which are inside the wall. Once the net is laid and both ends are on the bank then both ends of the net are slowly pulled ashore, making certain the lead line maintains contact with the bed and the cork line does not sink below the surface so that fish cannot escape. The main problem associated with seine netting from the shore is the need for a good bed profile over which to set and haul the net. Snags can prevent the net from being landed and silt and excessive weed growth can also prevent the seine net from working efficiently.

The main advantages of seine nets are that they are relatively cheap and easy to use. A trained and experienced operator can deploy a net in most circumstances and catch fish. The seine net will also catch a wide variety of the fish present, as it is the least selective of all the capture techniques.

Furthermore, seine netting, especially with micromesh nets, is often the only way to capture the smaller representatives of populations in shallow freshwaters.

Trawl nets

Trawl nets are similar to seine nets but are ' trawled ' behind a moving boat. Although more commonly associated with saltwater fishing, trawling can be successfully employed in surveying larger freshwater lakes. The major constraints are the cost, the difficulty of deriving quantitative assessments of the fish population under study and the damage trawl nets can do to the bottom.

Gill and trammel nets

Gill nets and trammels both work on the same principle of entangling the fish that swim into them. The gill net is a single piece of monofilament net suspended between a cork line and lead line. Fish swim into the net and the gauge is set so that it catches them just behind the head and the fish cannot escape. A trammel net is a modification of a gill net consisting of three layers of net. A slack, small mesh, inner panel of netting is sandwiched between two outer layers of netting, which are taut and have a larger mesh size. Again the fish are entangled within the netting.

The main problems with these types of net are that they are very selective for the size of fish, with mesh shape and material type also influencing the efficiency of capture. Gill nets do not allow the fish to be sampled and released afterwards, as they are almost invariably fatal to the fish caught. Even fish that survive initial capture are often so badly damaged that secondary infections of the wounds inflicted eventually kill the fish. Furthermore, there is a good chance of catching non-target species such as diving birds and mammals. Results for gill-nets are also difficult to interpret, especially for estimates of absolute abundance, and have to be used mainly as a relative method.

The main advantages for these methods are their ease of use.

Trapping

Traps fall into two broad categories: baited and unbaited. Bait is used to induce certain species to enter a trap and, as such, they are quite selective. Baited simple drum traps are mainly used as a means to capture species for food, or to remove nuisance organisms such as signal crayfish. Unbaited traps are used either to intercept migratory species, or species undergoing local movements. Such traps are often combined with leader nets or fences, the commonest example of this in the UK being the fyke net. Fyke nets can be used in still-waters where they consist of cylindrical or cone-shaped netting bags mounted on rings. These rings get smaller as you move down the net. Each has a tapered entry net, which allows entry into each section but no exit. They usually have wings or leaders, which guide the fish towards the entrance of the bags. Fyke nets are fixed on the bottom by anchors, ballast or stakes. The application of trapping methods for population assessments is limited by several factors, not least being that they are absolutely dependent on the behaviour of the fish in relation to the type of trap used. For example, fyke nets select only fish that are actively moving around and therefore susceptible to the trap. They are also mesh selective in that fish smaller than the size captured by the mesh of the leaders escape. Furthermore, they cannot be left unattended as they may catch non-target species such as otters or diving birds.

Fyke traps are relatively easy to use and can work well. For example when there is an individual target species and age structure, growth or condition are the assessment criteria. For example, the Lake Windermere perch (*Perca fluviatilis*) trap has been a valuable aid in the long-term study of the lake. Disregarding the difficulties of quantitative appraisal, traps remain popular for studying juvenile fish, which may be below the size that can be taken by other methods.

Electrofishing

The electrofishing survey technique uses electricity to stun the fish, allowing them to be retained to be weighed, measured, aged and then returned to the water.

The electrofishing machine creates an electric field in the water that interferes with the central nervous system of the fish. This stuns the fish and enables them to be scooped up in a hand net. Electrofishing can work effectively in a variety of habitats but it is considered best suited to sampling from flowing waters. However it can be used effectively on still-waters. The type and method of electrofishing depends on the size of the water and the water's ability to conduct electric current. Either alternating or direct current (AC or DC) is used, depending on circumstances and operated either from a boat or using a backpack unit. The DC current works best in waters with a high potential to conduct electricity (hard waters). AC current is most effective in waters with a lower potential to conduct electricity (soft waters). Where appropriate, DC current is the preferred type of current because it has the least potential to harm individual fish. Sampling with electrofish-ing gear is efficient, quick and relatively safe if properly conducted by trained personnel. For electrofishing, the effective fishing range from the probe (anode) depends on biological effects such as species and length of fish as well as physical effects like conductivity of water or substrate type.

The main problem with electrofishing is that it is selective. This is because different fish species are more susceptible to the electric current than others and the size of the fish is very important. The longer the fish then the greater the potential difference and the more effect the electrofishing has. Furthermore, in still-waters the depth of the water and the width is very important. If the water is too deep then the fish may just swim down away from the anodes and evade capture or they may move away from the direction of the boat and so again evade capture. Even so, it remains one of the more effective methods for sampling areas with a lot of weed growth and marginal areas. It is worth noting that electrofishing can do significant harm to fish if used incorrectly. If the current is set too high the fish muscles can spasm in such an extreme manner that the backbone is broken or the artery below the backbone is ruptured. It is difficult, if not impossible, to spot this damage to the fish when it occurs, so it is not apparent to the operator at the time. It is therefore very important to set the current as low as possible to catch the fish and to reduce the amount of time individual fish are in the electrical field.

Fishermen's catch surveys

The use of information from fishermen can give access to a vast amount of data, even if the primary reason the fish are caught is for leisure. Surveys based on fishermen ' s catch returns have the advantage of being able to run over much longer timescales than might be possible with scientific sampling programmes. There have been numerous studies using this form of data collection, which have proved extremely beneficial in examining long-term trends. For example, trends in species composition over an 11-year period in the lower reaches of the River Trent showed that the fishery changed from a roach/dace to a bream/chub community over a period of a few years. This type of data would be hard to substantiate without the involvement of the fishing community.

This method is also a useful tool when sampling small, still-water commercial fisheries, for example sampling during a competition match. When the fish are in the keepnets it allows for relatively easy representative population samples to be recorded after weighing. Scales can also be easily removed for the purposes of ageing. This method can be effective in gaining a relatively broad range of data in a short time- frame to give a ' snapshot ' of fish populations at a given time.

There are disadvantages with using catch data, however, the most notable being the dif-ficulty in estimating abundance. There will also be species bias and the quantification of effort, not to mention the fact that healthy fish will intrinsically feed more than unhealthy fish so large sections of the actual population may not be represented.

Non-capture methods

It is possible to gather some good information without actually catching or handling the fish. These are called non-capture techniques and involve some high-technology but very useful equipment.

Counters

This technique is used primarily for electronically counting migratory salmonids in rivers. As the fish pass over a structure with electrodes attached they are electronically counted. The technique can be very efficient, with over 90% of the fish passing over the structure being successfully counted. The method is only really applicable to migratory river species, but can be adapted to some still-water situations.

Hydroacoustics

Hydroacoustics is a general term for the study and application of sound in water, usually by means of sonar. The practice is used for the assessment and monitoring of underwater physical and biological characteristics. Sonar provides accurate underwater topographic information, which is valuable in population surveys, for example assessing where in the water column individual fish populations prefer.

When assessing fish numbers, however, the method relies on the sound signal being bounced back from the swimbladder of the fish. These signals are analysed using complex data analysis techniques, and confidence in the results can be limited. The advantage of the method is that it provides access to fish populations that cannot be easily sampled in other ways such as in large deep lakes. Sonar equipment can successfully count fish although it is recognised to work most effectively in deep, open water with few species present. Species and size separation is very problematic but the development of multiple beam systems is making progress in this area.

The technique uses either classical vertical sonar beaming, as in the sea, in deep lakes or reservoirs, or a new technique of horizontal beaming for rivers and shallow reservoirs.

The development of horizontal beaming techniques has demonstrated the power of scientific echo-sounding as an ecological tool for estimating the abundance and size structure of fish communities in large lowland rivers, for which no alternative survey technique of such power exists.

The output from one survey using an Eagle echo-sounder. This display shows the fish marked on the screen and the depth at which they were located.

This technologically advanced method is obviously prohibitively expensive in most private fisheries, although it is a valuable environmental tool. Although the research using this method is essential in advancing ecological survey techniques, fishery managers must rely on more practical methodologies.

Direct observation

The observer can either watch and count fish directly or use remote control camera systems. Fish shoals in shallow waters have been successfully counted from boats, and size information can also be easily obtained. This method is somewhat crude, however, with notable constraints such as water clarity, cost and training. There is scope to use the technique in certain situations, for example where more traditional sampling techniques are inefficient or result in unacceptable disturbance of the fish population. In still-water fishery situations, however, observation alone cannot be relied upon to give representative assessments of fish populations and certainly cannot provide more in-depth community -related data.

Further information

The main advantage of capture techniques over non-capture techniques is that they allow further data to be collected. Once a fish has been caught further information can be obtained including the following:

- length;
- weight;
- scale samples;
- clear identification of the species;
- health samples;
- reproductive state.

The use of this information will be discussed later in this chapter.

ESTIMATING A BUNDANCE

Mark-and-recapture estimates

The mark-and-recapture method requires that you accept that the following conditions have the same probability of occurring:

- marked and unmarked fish have the same mortality rates;
- marked and unmarked fish are equally vulnerable to capture;
- marks are retained during the sampling period and all marks on recaptured fish are recognised;
- marked fish randomly mix with unmarked fish;
- there is negligible emigration or immigration during the recapture period.

The general process for estimating the size of a fish population using the mark-and -recapture method entails the following:

I. collecting a sample of fish of the target species;
II. giving fish identifying marks, such as a tag or temporary fin clip;
III. releasing fish in good condition back into the same area;
IV. allowing at least 1 day for marked fish to recover and become mixed in the population;
V. collecting another random sample of fish during a subsequent recapture run;
VI. noting the number of marked fish and unmarked fish by species and size;
VII. calculating for each combination of species and size group (to compensate for sampling selectivity) an estimate of abundance using the equation below;
VIII. summing the size group estimates by species to obtain an estimate of the total population within the size range actually sampled.

Estimation the size of a fish population, N (and variance of N), can be obtained using the following equation:

$$N = (M + 1)(C + 1)/(R + 1)$$

where

C = total number of fish caught in second sample (including recaptures);

M = number of fish caught, marked and released in first sample;

R = number of recaptures in the second sample (fish marked and released in the first sample);

N = estimate of population size.

The mark-and-recapture method is a useful method and generally considered to be more accurate than most. However, it is expensive, time consuming and requires careful handling of the fish to mark them.

There are various tags available for use in fish. Visible implant tags are good examples of a product that is easy to use. An example is a soft Visible Implant Alphanumeric (VI Alpha) tag, which is a small fluorescent tag with an alphanumeric code designed to identify individual specimens. VI Alpha tags are implanted internally just under the skin usually around the head. They remain externally visible for easy identification and show up remarkably clearly under blue LED light. Visible implant elastomer tags are also available, which mark fish but do not have an individual number. These are cheaper and easier to use.

It is not always necessary to use complex tags. It is usually sufficient for mark-and-recapture experiments to mark the fish by simple fin clipping or by injecting a dye.

Depletion estimates

Fish capture techniques also allow abundance estimations to be made through the use of depletion surveys. These generally rely on the concept of a constant capture effort being applied to the same population, the size of which declines as the fish are removed. This should cause the catch per unit effort to fall as it gets more difficult to catch the remaining fish. Extrapolation from a depletion or removal rate is therefore possible with relatively straightforward calculations.

A constant effort, using a selected capture technique, is applied in a defined area, on at least two occasions. The fish caught on the first occasion are retained until after the second capture is complete. The anticipated decline in catch per unit effort is used to calculate (by extrapolation) the initial absolute number of fish present.

The depletion method is satisfactory, although realistically all the data for one site must be collected in a single day. It also works better if the population being estimated is relatively small (say fewer than 2000 individuals).

This method requires that an adequate number of fish be removed on each sampling pass so that measurably fewer fish are available for capture and removal on a subsequent pass. Two types of depletion method are used: two-pass and multiple-pass. Because of differences in sampling selectivity, an estimate, by species and size groups, is normal. For both two-pass and multiple-pass methods, size group estimates and their variances are summed, as with mark-and-recapture methods, to provide estimates of the total population. The following conditions must be met for accurate depletion method estimates:

- Collection effort and conditions, which affect collection efficiency, such as water clarity, must remain constant.
- All fish within a specified sample group must be equally vulnerable to capture during a pass.

- Vulnerability to capture of fish in a specified sample group must remain constant for each pass (e.g., fish do not become more wary of capture).
- Emigration and immigration by fish during the sampling period must be negligible.

Depletion estimates are made using the following the general process:

I. Complete capture operation within a discrete section of water body.
II. Record number of fish removed, by species and size group.
III. Do not return fish to the water body.
IV. Repeat steps 1 and 2.
V. If steps 1 and 2 were completed twice, calculate population estimates using two-pass equations.
VI. If steps 1 and 2 were completed more than twice, calculate population estimates using multiple-pass equations.

Two-pass depletion methods

For two-pass depletion estimates, fish are captured and removed during two capture sessions.

Population estimate *N* and variance of *N* are calculated as follows:

$N = C_1^2/(C_1-C_2)$

where

$C\,1$ = number of fish caught in first sample; $C\,2$ = number of fish caught in second sample.

Two-pass depletion estimates are quite unreliable when fewer than 20% of the population is caught per pass.

Multiple-pass depletion methods

These require three or more catch runs to be completed and rely heavily on consistent catch-ability of the fish. Although multiple-pass depletion methods are more accurate than two -pass depletion surveys, they require a consistency of effort that is difficult to achieve and for the fish to not show any wariness to the capture method being used. The statistical analysis of the catch data is complex and the requirements of this method are so exacting as to make its use inadvisable.

Population structure

It is normal when capturing fish for a population estimate to take additional samples that can be used to gain further information about the population being surveyed. This additional data may include length, weight, reproductive state, photographs, definitive identifications, scales and health samples. This information can then be used to give a variety of useful outputs.

Length–frequency analysis

Length is easy to measure and there is a broad correlation between length and age.

Once all the fish (or if there are large numbers, a suitable sub-sample) are measured, then a simple graph of the length against the numbers caught can give you some very useful information. An example of a length – frequency histogram is given in Appendix C at the end of the book.

In the example represented by the graph in Appendix C, the first peak relates to one age group (year class) (2-year-olds) at about 50 mm long, then there is another peak at 70 mm long, which could represent a 3-year-old year class of fish. So in this example it is possible, with representative data, to see the distribution of year-class groups of young fish

up to three years. However, variations in growth between individual fish of the same age mean that this method of ageing is unreliable, except for the very youngest age groups. The rest of the graph, as expected, shows a fairly steady decline in numbers, as the fish get older. This sort of graph provides very useful information about the population structure and helps to indicate things like poor year classes, which are under-represented in the population. In some cases the decline in numbers can also be used to calculate mortality. The number of year classes or cohorts present in a sample depends on several factors:

- species longevity: this can range from one year in some ' annual ' species to several decades for long-lived species;
- prevailing rate of mortality, which varies with time and place;
- sample size and the selectivity of the sampling technique and pattern of sampling.

As a generalisation, more young fish than old can be expected, but population age structure will also be influenced by year-class strengths. The selectivity of the sampling method will generate differences between population and sample-age structure.

To make this information more useful, it can be validated by ageing some of the fish accurately.

How to age fish

There are two methods used to determine the ages of fish in a population. These are mark – recapture and use of growth checks on skeletal structures.

Mark–recapture

Fish can be marked, and if the mark is representative for that year class then the age from that year can be recorded and will be known and established when the fish are recaptured. This can provide direct data on growth and mortality over a very precise time interval.

Although a whole part of the life history is known, the age at first capture is often not known. Although this method is very accurate it is very time consuming, and tagging methods are often only visible for months rather than for years.

Use of growth checks on skeletal structures

This can provide a precise absolute age (in years at least) for all individuals in a sample. The species captured are identified and measured, and two to four removed, usually from the shoulder region. Otoliths (ear bones), fin rays and gill opercula can all be used to age fish, but their use means the fish must be killed, so scales are preferred. Any scale removed is regenerated by the fish. A few fish species do not have scales suitable for ageing (e.g., eels, tench, pike and perch). If these require ageing, otoliths are the preferred method. The principle behind ageing fish relies on the fact that many of the body structures, such as scales, have growth rings (circuli) laid down as the structure grows. These rings are calcium based and are probably for support and strength. The growing seasons can be seen on these structures as widely spaced growth rings. When the fish stops growing, for example in winter, the growth rings are not so widely spaced. With the onset of the new growing season the circuli become much more widely spaced again and a mark (annulus), representative of a year, is formed. During the period of no growth it is not unusual for the edge of the scale to be eroded, so when new growth starts it begins on an eroded edge and the new circuli may be overlaid across partly eroded ones. This is a clear indication of an annual mark. The number of annuli is equal to the age of the fish.

It is very clear where the annual marks are. For many species this is true and it is relatively straightforward to tell the ages, particularly of young and fast-growing fish. However, some species such as tench, pike and perch are almost impossible to age from scales. Furthermore, the older a fish gets the more difficult it is to age. For the first few years of life fish tend to grow quickly, but as they mature they put more of their energy into producing gonads and so grow less quickly. As the annuli get closer together as they get older, they become increasingly difficult to tell apart.

Information gained from ageing fish

The age of fish can give you a lot of other useful information including the following:

- establishing the longevity and individual life history of the fish;
- defining population structure;

- establishing individual and population growth rates;
- determining age and size at maturity.

Knowing the age of each fish and the length at which they were caught gives the average length at age.

Age group notation

In the UK (which is in the temperate northern region) the growth season is regarded as 1 April to 30 September. For fish sampled in this period, growth is still visible by following the annuli formed the previous winter. This is termed ' + growth ' . It is anticipated that the growing season will become extended as the predicted climate change occurs. Fish that are sampled in the winter months are in a period of little or no growth and the annulus is formed. The edge of the scale is therefore the end of the previous growing season and the age is given in full years (' no + growth ').

ESTIMATING HEALTH

The health of a fish, and therefore the population, is obviously useful information. Some information on the health of the fish population will be apparent from the result of a fish population analysis. Also, ageing the fish will give information on the growth rate of the fish in the population and age at maturity, again useful indications of the overall health of the population.

How to obtain health information from live and dead fish, but other parameters can indicate the health of the stock as a whole. For example, if length and weight were both recorded during the initial data collection exercise, the ' condition factor ' of the fish can be worked out using a simple formula. This is a mathematical formula for determining the physiological state of a fish. It is calculated by dividing fish weight by length cubed (W / L^3). The heavier a fish for a given length, the higher its condition factor (K). The higher the condition factor is, the healthier the fish is.

However, the overall health of the fish in terms of their disease status is important as well. Knowledge of the pathogens found in a fishery is important for the following reasons:

- in the event of a mortality problem, a ' normal ' level of pathogens is known;
- it may allow any potential disease risks to be managed before they become a problem;
- it gives information on the overall health and stability of the fish populations;
- it may inform other fisheries management options, such as adding or removing stock.

6

Fisheries and Marine Biodiversity

ADAPTIVE CAPACITY IN MARINE RESOURCE GOVERNANCE

Marine resource governance is a global challenge of increasing magnitude and importance. Depletion of fisheries resources, coastal habitat degradation, threats of global warming, food security at risk and rising fuel costs all contribute to the challenge. Small island developing states (SIDS) in the insular Caribbean are particularly vulnerable to natural and anthropogenic threats. Fishing and coastal communities and enterprises depend heavily on marine resources and ecosystems. Cross-scale and multi-level networks for marine resource governance may provide means for enhancing resilience and adaptive capacity in these complex social-ecological systems. Understanding how local, national and transboundary networks link actors at various levels and scales across the region in their quest to use and manage shared marine resources is key to developing policies and practices that enable good governance. The nature of such networks is being investigated through cases in small-scale fisheries governance in order to determine what features, including collaborative processes, relate to resilience and the capacity to adapt. This research is a work in progress, the conceptual framework and preliminary findings of which are shared in this paper, along with early ideas on how research outputs can be communicated and used to improve marine resource governance in the Caribbean.

The opening sentences of the Caribbean Sea Ecosystem Assessment (CARSEA), a sub-global component of the Millennium Ecosystem Assessment (MA) proclaim that: "The peoples of the Caribbean are defined by the Sea whose shores they inhabit. In the rich diversity of cultures and nations making up the region, the one uniting factor is the marine ecosystem on which each ultimately depends" (CARSEA 2007: xiv). Yet, Caribbean marine resources are in trouble, as is the case in many coastal and marine areas worldwide. Depletion of fisheries resources, coastal habitat degradation, global warming, food security at risk and rising fuel costs all contribute to the challenges faced by the small island developing states (SIDS) in the insular Caribbean. These places and their people, characterised by coastal communities and enterprises (mainly fishing and tourism) heavily dependent on marine ecosystems, are particularly vulnerable to both natural and anthropogenic threats to life and livelihood. Member states of the Caribbean Community (CARICOM) are among the most numerous, dependent and vulnerable in the region. Looking closer at fisheries, throughout CARICOM there is a need to develop alternatives to conventional fisheries management, paying particular attention to people-centred approaches and governance.

On a global scale conventional science-centred approaches to fisheries management have not performed well, and some scholars have deemed them to be failures (Berkes et al. 2001). As a result, fisheries have long been said to be in crisis (McGoodwin 1990) despite recent attempts to incorporate ecosystem approaches to fisheries (EAF) (FAO 2003, Christie et al. 2007) and add human dimensions to this (De Young et al. 2008). Greater attention not only to ecosystem approaches to marine science, but to governance as well, is needed to address situations that threaten Caribbean social-ecological systems. This is the convergent conclusion that can be drawn from several sources as summarised in CARSEA (2007) and the Caribbean Large Marine Ecosystem (CLME) project (CLME 2007). These reports, and this paper, focus upon the use of living marine resources and governance of small-scale fisheries within the context of ecosystem approaches.

Fisheries management and integrated coastal management are still relatively new in CARICOM (Breton et al. 2006, Mahon and McConney 2004). Before conventional approaches firmly take root, the opportunity exists to investigate, test and propose approaches based on concepts and models drawn from complex adaptive system (CAS) and social-ecological system (SES) perspectives to determine whether these new alternative directions offer greater promise of success than the conventional (CERMES 2007). In the fisheries literature these perspectives have not yet become mainstream, but there is a growing interest in their theory and practical application, especially with regard to marine resource governance in small-scale fisheries (Bavinck et al. 2005, Kooiman et al. 2005, Mahon et al. 2008).

At the core of the new perspectives are concepts and practices still unfamiliar to many policy decision-makers, management authorities, resource users and other stakeholders engaged in marine resource governance. These core concepts include social networks, resilience and adaptive capacity. Put simply, they are related in that network structures and relationships provide a useful way to represent complex SES, resilience can be a desirable feature of SES, and adaptive capacity is one of the means of achieving and sustaining resilience in SES. We are engaged in research and development to put some of this thinking into practice (CERMES 2007).

In this paper we examine small-scale fisheries governance in the CARICOM region with an emphasis on the role of user groups, particularly fisher folk organisations (FFOs). This considers what and how multi-stakeholder collaborative planning that links fisheries authorities and fisher folk at national and regional scales with university researchers can contribute to resilience thinking and action. At this early stage of engagement much depends upon communication, so we also pay attention to how information exchange is helping to shape this ongoing experiment.

Caribbean marine resource governance

The Caribbean is culturally, ecologically, geographically, institutionally, politically, socially and otherwise diverse and complex (Chakalall et al. 2007). Considering only the 15 CARICOM Member States scattered across the region does not significantly reduce complexity.

From an ecological perspective, one of the main contributors to complexity is that most marine resources, both exploited and non-exploited, are shared across several maritime jurisdictions at various points in their life cycles. From a social system perspective, there is also considerable interaction of people and issues across jurisdictions. Marine resource governance in the region must therefore be transboundary, crossing the Caribbean at multiple scales and levels. Some transboundary linkages are not commonly known, and consequently the need for collaboration in governance is poorly appreciated (Chakalall et al. 1998, 2007). This SES cross-scale complexity and limited collaboration in governance results in what some authors refer to as a maze of regional and sub-regional institutions, such that the many overlapping, competing arrangements for various aspects of marine governance do not provide a coherent governance framework (Chakalall et al. 2007).

This is a critical problem, especially given the limited state and civil society capacity for fisheries management within the region, and the scarcity of fisheries officers trained in social sciences and related skills (Mahon and McConney 2004). To reduce the problems of scale mis-match (Cumming et al. 2006), while not discounting the larger Wider Caribbean complexity and the several other regional initiatives that concern marine resource governance such as the CLME project (Fanning et al. 2007), we concentrate on arrangements linked to the Caribbean Regional Fisheries Mechanism (CRFM) established in 2002 by intergovernmental agreement as a body of CARICOM. The early evolution of the CRFM is described by Haughton et al. (2004). The mission of the CRFM is "to promote sustainable use of fisheries and aquaculture resources in and among member states by the development, management and conservation of these resources *in collaboration with stakeholders* to benefit the people of the Caribbean region" (emphasis mine).

According to its web site (www.caricom-fisheries.com accessed 27 Sep 08), "the CRFM is the core of a *complex interactive network* of a wide variety of stakeholders in fisheries" (emphasis mine). Three organs together make up the CRFM: (1) the Ministerial Council or Body; (2) the Caribbean Fisheries Forum; and (3) the CRFM Secretariat. The former makes policy decisions and the latter provides technical and administrative support. However, for fisheries stakeholders, most of the action should take place in the Forum.

The Forum is the main technical and scientific decision-making arena, but it may also be the most important in terms of fisheries governance unless its advice is questioned or overridden by the Ministerial Council. The Forum comprises full members, associate members, and observers. The web site says that "Observers may include, but are not limited to, intergovernmental fisheries organisations, fisher folk organisations and other NGOs, and universities". The CRFM Secretariat is the secretariat of the Forum.

Based on the foregoing, the Forum is an attractive target for FFOs and private sector fishing industry groups that wish to get into the CARICOM policy domain through the route of being observers. It is also possible that they may be included in member state delegations as has been the case with the Suriname Seafood Association, but this is less likely in most other member states. Apart from the Suriname example, FFOs have been absent from the Forum and none are accredited to it as observers on their own. Another attraction of the Forum is that in 2003 it formed an open-ended working group to take on the task, directed by the CARICOM Heads of Government, to establish a Common Fisheries Policy and Regime (CFPR). The CFPR is still under negotiation. What its final form and function will eventually be is presently unpredictable. However, the CFPR may become one of the main instruments for shaping and implementing fisheries governance within CARICOM and perhaps beyond. Early drafts of the negotiation text suggest that the Forum contemplates having, through the CFPR and otherwise, a wide range of decision-making responsibility and authority. Matters for its consideration may range from all aspects of harvest operations (including perhaps access and allocation), to postharvest and trade, marine aquaculture, research, enforcement, dispute settlement and more (Phillips et al. 2006).

One of the main reasons that FFOs have been unable to engage the CRFM in marine resource governance is that in most countries they lack the capacity at both local and national levels, and they have not yet formed a regional level organisation. McConney (2007) describes how FFOs were introduced to many locations in the English-speaking Caribbean (now the Member States of CARICOM) during the British colonial period, often in the early 1960s and

1970s. The histories of how they rose, and often fell, at different times in different places vary, but common inter-related features of their state-directed formation typically include:

- Being part of the movement towards independence
- Use for political empowerment of "the small man"
- A means to alleviate poverty and encourage saving
- For pooling money to improve commercial business
- An emphasis on boat owners rather than fishers
- Channelling of government subsidies for fishing
- Efficient supply of inputs for fishery development
- Getting into bigger, better and costlier boats
- Seafood processing for food security and export
- Starting with savings societies and cooperatives

This illustrates a variety of reasons behind organising, and of different types of organisation (harvest, gear and fuel supply, processing, etc.). Early cooperatives failed in many places after only a few years for reasons including:

- Under-capitalisation (not enough funds to run them)
- Limited capacity (leadership, skills, resources, etc.)
- Inadequate support from government agencies·

Membership bases that were too small to be viable

- Little succession planning to groom new leaders
- No follow-up to projects and technical assistance
- Short term planning rather than strategic planning
- Limited oversight and monitoring to give guidance
- Financial mismanagement ("who thief the money?")
- Regulatory agencies did not correct financial wrongs
- Connection to political movements that changed
- Inability to sustain collective action outside of crises
- Government programmes that competed with them

Although the development trajectories of FFOs in CARICOM are poorly documented, through oral history we are able to look back on these past experiences and learn from them.

Yet, learning from past successes and failures (e.g. Meynell 1984, 1990) is not adequate preparation for the present and future if situations have changed (Pollnac 1988). How things have changed varies by location and the diverse nature of Caribbean fisheries SES, but the change that most concerns us here is the recent emergence of the CRFM and the opportunity for involving FFOs in cross-scale governance at integrated multiple levels whereas previous efforts were mainly local or national (McConney et al. 1998, McConney 1999, Brown and Pomeroy 1999).

Recognising this opportunity, the CRFM initiated a project on the 'Development of a *Caribbean Network* of Fisher Folk Organizations' (emphasis mine) in September 2006, implemented through its Secretariat. The demand for such a project was based on the results of a regional needs assessment (CRFM 2004) and subsequent regional meetings in 2004 and 2005 that involved FFO representatives and recommended the formation of a regional network of national fisher folk organizations (NFOs). Strengthening FFO capacities was to be aimed at addressing:

- Lack of critical mass of NFOs for launching the regional network
- Fisher folk organizations' weak management skills
- Insufficient access to relevant information by fisher folk

The 2005 meeting produced a 'Strategy and Medium Term Action Plan for the Institutional Strengthening of Regional Fisher Folk Organizations-2006 to 2010' to address the points above. It is through this bottom-up collaborative planning process that the project was designed. The specific purpose of the project is to have "institutional capacities of fisher folk organizations developed at the regional, national and community levels" (CRFM 2007). In the two years since its inception in 2006 the 'FFO project', as it has come to be known, has held three regional training workshops, created a directory of fisheries stakeholders, produced bi-monthly newsletters, spawned a fisher folk lead coordinating unit (CU), sponsored exchange visits of fisher folk leaders and supported the implementation of a work plan to establish the networked regional fisher folk organisation (RFO).

The structure and functions of the latter are still being worked upon. Researchers at the University of the West Indies (UWI) Centre for Resource Management and Environmental Studies (CERMES) have been involved in the FFO project as a partnership with the CERMES project on 'Marine Resource Governance in the Eastern Caribbean' (MarGov) Project. The latter is participatory action research aimed at developing capacity for marine resource governance. The goal of MarGov is to understand marine resource governance related to small-scale fisheries and coastal management in the eastern Caribbean using CAS and SES concepts (CERMES 2007). Before we examine the relevance of the FFO project to this, the use of some concepts is briefly explained.

THE GLOBAL IMPORTANCE OF FISHERIES AND AQUACULTURE

Fisheries play an important role in the world food economy. Fisheries are a source of employment for about 200 million people who depend directly upon ocean fishing for their livelihoods. Fish is the primary source of protein for some 950 million people worldwide and represents an important part of the diet of many more. In less than 50 years, the world's average per capita consumption of fish has almost doubled. Globally, fish provide about 16% of the animal protein consumed by humans, and are a valuable source of minerals and essential fatty acids. Fish is the primary source of omega-3 fatty acids in the human diet. Omega-3 fatty acids are critical nutrients for normal brain and eye development of infants, and have preventative roles in a number of human illnesses, such as cardiovascular disease, lupus, depression and other mental illnesses.

The reported production of fish for direct human consumption doubled between 1950 and 1970, and has stabilized since then at an average of 9.0 to 10 kg of fish per capita, notwithstanding world population growth. Fish consumption per person is expected to continue to rise. Supply will probably be limited by environmental factors, and a likely range for demand is 150 to 160 million tons, or between 19 and 20 kg per person in 2030. Global increases in consumption of food fish will take place predominantly in the developing countries, where population is growing and higher incomes are allowing purchase of high value fisheries items for the first time by many people. However, fish production in least developed countries where fish protein is needed to prevent malnutrition is a key element of food security in these regions and a critical area where innovative programs are needed to increase production.

Today, fishing is the largest extractive use of wildlife in the world. The value of world total fishery production in 1999 was US$ 125 billion. World production of fish, crustaceans, mollusks and plants reached 142 million tons in 2001. Capture fisheries production, which accounted for 66% of the total, was 93.7 million tons, of which inland capture was 8.7 million tons, while aquaculture production was 48.4 million tons including plants. Marine and freshwater fish are also an increasingly important recreational resource, both for active users such as anglers and for passive users such as tourists, sports divers and nature-lovers.

Asia predominates in capture fisheries and aquaculture production, with China the leading nation worldwide. In 1999, over three-quarters (or 97 million tons) of the global production of fish, crustaceans and mollusks were utilized for direct human consumption. Fish utilized as raw material for the production of animal feed, e.g., fish meal and fish oil, represented about one-quarter of the total fishery production in 1999; and this amount has remained relatively static for the past 15 years. Most growth in the fisheries sectors is projected to occur in developing countries, which will account for 79% of food fish production in 2020.

Increased production of fish from aquaculture has occurred primarily as a result of increasing feed inputs into ponds and other production systems, thereby increasing yields per hectare by an order of magnitude compared to extensive production systems in which rearing water is fertilized only. Higher inputs mean two things to the aquaculture feed industry; more feed and higher quality feed. Currently, global feed production for farmed fish and crustaceans is approximately 13 million tons, and predictions are for feed production to increase to over 37 million tons by the end of the decade, an increase of 24 million tons. Feeds for salmonids and marine fish have always been complete feeds, i.e., ones that supply all of the nutritional needs of the fish. Pond-reared fish, in contrast, obtain a significant proportion of their nutritional needs from pond biota. The degree to which feeds must supply essential nutrients to pond-reared fish increases as rearing densities increase beyond the capacity of natural foods in ponds to supply them. Fish farmers around the world have found that as they increase feed inputs, the biomass and economic yields from ponds increase as well. Thus, great areas of low-input, pond-based aquaculture mainly in Southeast Asia and China are being converted from low-input systems to high-input systems that depend upon high quality feeds to supply an increasing proportion of nutrients used by the fish.

The effects of the aquaculture industry's growth and of changes in feed input in pond-based aquaculture have been dramatic with respect to the use of marine proteins by the aquaculture feed industry. In the mid-1980s, less than 10% of annual fish meal production was used by the aquaculture feed sector. Today, that proportion is over 40%. Similarly, aquaculture now uses nearly 75% of annual global fish oil production, up from less than 10% seventeen years ago, but this chage is mainly due to the adoption of high lipid feeds by the salmon farming industry, rather than increasing the inputs into pond-based aquaculture systems.

Fish meal and oil are produced from species of fish that are not generally utilized directly for human food, e.g., herring and capelin in Norway and Iceland, sand eel in Denmark, capelin in South Africa, anchovies in Peru and northern Chile, jack mackerel in central Chile, sardines in Japan, and menhaden in the USA. The rapid increase in fish meal and oil use in aquaculture feeds over the past 15 years has not resulted in over-exploitation of the stocks of fish harvested primarily to produce fish meal and oil. Production of fish meal has averaged between 6-7 million tons per year since the mid-1980s, except in El Nino years, when production has been lower. Fish oil production has averaged between 1.2-1.3 million tons. No significant changes in annual harvest or fish meal/oil production associated with increased aquaculture production or increasing intensification of aquaculture and concomitant higher feed inputs is evident. Increased fish meal/oil use in the aquaculture sector has come at the expense of other uses.

Estimates of future protein requirements for aquaculture feeds depend upon future production from various segments of the aquaculture industry and on annual production levels of fish meal. As aquaculture production expands, it will be essential to replace portions of fish meal and oil in fish feed formulations with alternative ingredients derived from alternative sources, primarily grains and oilseed products.

Because 70% of the fish meal used in diets for fish and crustaceans is used to produce diets for salmonids, marine fish and shrimp, these production sectors are the focus of most research with respect to the use of alternative protein sources. Numerous studies have been conducted to evaluate the effects of replacing various percentages of fish meal in diets for these fish, and, without exception, none has successfully replaced 100% of fish meal without reducing fish performance. At best, 50% of fish meal in diets for salmon and trout can be replaced by soy protein concentrate, and 25-30% with soybean meal. Similar findings have been reported in studies of wheat gluten meal, corn gluten meal, and rapeseed protein concentrate. For the most part, formulated fish diets are used to produce high-value fish for export. In underdeveloped countries, increased production of lower-value fish that are consumed by local populations will depend upon the use locally-available, low-cost feed ingredients that can be combined to produce prepared feeds to increase productivity of community ponds and waterways.

This is an area that requires substantial investigation and development, and is a logical target for USAID support. Another logical target for USAID efforts is the development and testing of feeds that are based upon sustainable plant-derived protein sources for use in developing countries where fish are grown both for domestic consumption and export, such as China. Partnerships with US commodity groups could be a mechanism to extend USAID efforts. It is critical that USAID become a leader in this area, given the perception that increased aquaculture production leads to higher fish meal use and greater pressure on stocks of fish that are captured to produce fish meals and oils.

Fisheries products have become the most internationally traded food, as some 37% (by quantity) of all fish for human consumption is traded across borders. In 1999, international trade (in live weight equivalent) represented 34% of the total production. In 1999, foreign trade earnings amounted to US$ 52.9 billion. Most fishery exports were destined to developed countries, with Japan the world's leading importer, and the USA second. The US balance of trade deficit in seafood products is approximately $9 billion.

Exports of fisheries products to the developed countries have become so lucrative that nations like Argentina, a traditional, globally important exporter of meat products and livestock have turned to seafood exports as the major source of foreign exchange earnings. In 2002, exports of raw and processed fish and shellfish from Argentina surpassed meat products and livestock, with beef earning US$ 574 million in export revenue, while seafood—mostly prepared products—earned US$ 714 million. Argentina's main consumers of domestic seafood products were Spain, Brazil and the US.

NEEDS FOR ENHANCED MANAGEMENT & EXTENSION CAPABILITIES

Human resource development and institutional strengthening are widely held to be the principal requirements for improving integration at the level of individual farms and communities, in river basin and coastal area management, and at the level of sectoral and macroeconomic policies. Co-management and community-based management approaches to the use of common-property resources have received increasing attention in recent years because of their assumed greater efficiency and prevention of undesired distributional implications.

Land use planning and zoning, together with environmental impact assessment procedures, are vital tools for minimizing conflicts between resource users, negative environmental impacts and enhancing sustainable development. The involvement of fisheries agencies in these activities therefore is clearly essential. Factors that the users themselves identify as being important for successful resource management include: small group size (which facilitates the formulation, observance and monitoring of a collective agreement), social cohesion, resource characteristics that facilitate the exclusion of outsiders, and visible signs of successful collective management. At the level of river basins and coastal areas, integration is aimed at managing components as parts of a functional whole, explicitly recognizing that management needs to focus on human behavior, not physical stocks of natural resources such as fish, land or water.

Technical adaptations alone are inadequate to direct sustainability of complex enterprises such as aquaculture and fisheries. An improved, and more participatory, applied research process can stimulate a greater momentum for change and increase the effectiveness of aquaculture/fisheries extension approaches. In order to ensure the ecological and social sustainability of fisheries and aquaculture systems, new methodologies must become more widely available to capture both the global (macro) and the fishing/farm-level (micro) social ecological processes occurring in order to determine the appropriate paths for research and extension interventions. As an example of what is required, the USAID Pond Dynamics/Aquaculture CRSP conducted groundbreaking micro-economic analyses of markets in Central America which provided invaluable data to support decision-making on pond aquaculture facilities. Such efforts need greater emphasis and need to be expanded to include analyses of nearshore and inland capture fisheries issues. In addition, much more attention needs to be given to marketing and market development issues. Small scale fishers and farmers in developing countries require additional education and training to help them determine markets capacities for their products. Larger producers need additional education and technical assistance with marketing plans that include basic information on market networks.

Participatory technology development (PTD), pioneered by the workers in the fields of farming systems research and extension and agroecology, has been found to be a very successful approach when applied to enhance the sustainability of fisheries and aquaculture farming systems.

PTD has been called "participatory learning for action". PTD uses assessment tools to enable fishers and farmers to analyze their own social ecological situation and to develop a common perspective on natural resources management and fisheries/food production at the local level. PTD involves group facilitation methods, methods for interviewing and dialogue, and visualization/diagramming methods.

In farming and fishing systems PTD approaches the natural and social ecological sciences come together in a very unique form of interdisciplinary environmental scholarship. Fixed answers derived from fisheries/agriculture/aquaculture experiment stations and passed down to farmers in a top-down manner are too inflexible and cannot solve the problems of sustainability which require more site specific, integrated social and ecological methodologies. Aquaculture farmers and coastal fishers are innovative groups in their communities, and farmers/fishers oftentimes know as much as researchers—making the reversal of traditional extension roles a given which should be embraced as a set of unique opportunities for collaborative work—not challenged or fought. By initiating a new era of cooperative research with fishers/farmers, a more detailed and very intimate knowledge of the natural and social ecology of farming/fishing systems can be combined with useful scientific knowledge to evolve sustainable aquaculture and fisheries ecosystems.

MARINE FISHERIES RESOURCES AND POTENTIAL

Reported global production of marine capture fisheries increased from 17 million tons in 1950 to about 80 million tons in the mid-1980s, oscillating since then between 78 and 88 million tons (excluding discards), representing 60% of the overall fisheries production including aquaculture in 2001. The annual rate of increase of marine catches decreased to almost zero in the 1990s, indicating that the world oceans have reached their maximal production under the present fishing regime.

An estimated 25% of the major marine fish stocks for which information is available are underexploited or moderately exploited. Stocks or species groups in this category represent the main source for the potential expansion of total marine catches. About 47% of the main stocks or species groups are fully exploited and are therefore producing catches that have reached, or are very close to, their maximum sustainable limits. Thus, nearly half of world marine stocks offer no reasonable expectations for further expansion.

Another 18% of stocks or species groups are reported as overexploited. Prospects for expansion or increased production from these stocks are negligible; and there is an increasing likelihood that stocks will decline further and catches will decrease, unless remedial management action is taken to reduce overfishing. The remaining 10% of stocks have become significantly depleted, or are recovering from depletion and are far less productive than they used to be (or than they could be) if management can return them to the higher abundance levels commensurate with their pre-depletion catch levels. Recovery usually implies drastic and long-lasting reductions in fishing pressure and/or the adoption of other management measures to remove conditions that contributed to the stock's overexploitation and depletion. In most areas, overfishing is certainly a significant factor responsible for the declines. The information available tends to confirm the estimates made by FAO in the early seventies that the global potential for marine fisheries is about 100 million tons of which only 80 million tons were probably achievable for practical reasons. It also confirms that despite local differences, overall, this limit has been reached.

Developed countries are faced with fully or overexploited stocks so their management objectives concentrate on stock rebuilding and capacity reduction, although most countries also have significant aims regarding markets and social conflict. The most urgent objective is to scale fleet sizes so that they become commensurate with sustainable exploitation of the resources. Management plans also increasingly recognize the need for a policy that integrates fisheries with management of the coastal zone or inland waters.

Developing countries tend to concentrate on fisheries development in terms of new resources and technology. Although it is recognized that some stocks are overfished, objectives are concentrated more on enhancing and diversifying fisheries rather than on limiting fishing efforts. This is perhaps because the underlying concern for many countries is the relatively important role fisheries play in employment and food security for some of their poorest people. More specific aims include building infrastructure (particularly for processing to reduce post-harvest losses and increase the value added); fishery enhancement, through restocking; and reducing social conflicts, not only among different fishing groups but also between fisheries and other sectors.

The principal policy challenge is to bring the capacity of the global fishing fleet back to a level at which fish stocks can be sustainably harvested. Fisheries based on clearly defined rights of access will need to become more common: experience shows that when these rights are not merely in place but are understood and observed by users, conflicts tend to be minimized. The need is to blend innovation, research, conservation and educational awareness into a goal of aquatic sustainability, and to demonstrate this via successful commercial enterprises.

The "great trends" of the past 20 years in marine fisheries worldwide are the: (1) globalization of both production and labour in marine fisheries, (2) the overcapacity of fleets both nearshore and offshore, resulting in dramatic and widespread declines in catches per unit efforts, (3) the movement of the bulk of the world's fishing capacity from the developed to developing countries, resulting in (4) declining catches and economic hardships worldwide in small scale artesanal and inshore fisheries resulting in less fish for poor consumers, and (5) infant success with various property rights schemes and use of protected areas to sustain stocks.

Inland Fisheries Resources and Potential

Four current strategies in the use of inland waters for fisheries can be distinguished: (1) food fisheries based on wild stocks; (2) enhancement of food fisheries in smaller water bodies and reservoirs; (3) recreational fisheries, which are becoming more common in many areas of the world, and, where they develop, tend to supplant commercial food fisheries; and (4) locally very intense exploitation of juvenile or small adults for stocking into other water bodies and/or aquaculture ponds, or for the ornamental fish trade.

In 2001, inland capture from 129 countries reached 8.7 million tons. Regionally, Asia dominates inland capture, accounting for nearly 66% (about 5.8 million tons) of the total in 2001, with Africa second at 24% (nearly 2 million tons). In comparison, the remaining regions are relatively insignificant. The capture from inland waters is very diverse. Apart from fish, other groups included in inland capture that are tabulated by weight are freshwater shrimps and prawns, frogs, terrapins, turtles, crayfishes, mussels, mussel shells, pearl oyster shells, swamp crabs, marine worms, manatees and green seaweeds.

There is an enigma concerning the status of inland fishery resources. On the one hand, it is widely accepted and well documented that freshwater resources and environments are deteriorating rapidly and widely. On the other hand, harvests from freshwater environments are stable or on the upswing among countries that account for 93% of the total inland capture. The increases are due to two factors: (1) enrichment of aquatic systems from land based human activities that until now have been a counterweight to habitat losses and pollution; (2) enhancements of inland fisheries (e.g., stocking, introductions) that increase the output per unit of area; and (3) the increased number of reservoirs with capture fisheries enhancement activities. Inland fisheries enhancement methods, often combined with conventional fisheries management practices, are widely applied, and are becoming a central theme in the management of inland waters in developing countries The exact contribution of enhancements to the total inland capture fisheries production is, however, difficult to estimate.

About two-thirds of inland capture is from developing countries and about one-fourth is from China. The countries in Africa and Asia that are among the top 20 in inland capture production also rank low in the Human Development Indices. This comparison underlines the need to sustain inland fisheries in the countries where inland capture is relatively important in its own right, where inland capture contributes significantly to food security, and where levels of human development are lowest. In contrast, inland recreational fisheries are pre-eminent in most developed countries; however, almost every country with inland waters has some recreational fisheries that often play an important role in subsistence. Indications are that freshwater recreational effort is perhaps one-half of the food fishing effort from a worldwide perspective.

Output from inland food fisheries will continue to increase slowly overall. In the short term most of the gains will be realized where enhancements are already common, namely in Asia. In Africa, where enhancements have been slow to be implemented, output may actually decrease due to overfishing and a deteriorating environment in the most populous countries, but increase as enhancements become more widely implemented. Likewise, in Latin America, inland fisheries enhancements are practiced and are growing, but the demand for inland fish is weak in many countries, while in others degradation of the environment is causing a decline in production.

Degradation of the environment is the main underlying issue, and the consequent loss of fishery habitat is the pre-eminent concern. A closely related concern is that loss of habitat that, along with the intense and widespread exploitation of fishery resources, negatively impacts aquatic biodiversity. In practical terms, the resiliency of resources to fishing, and often the quality of the resources, is lessened. Additionally, in most countries the main challenges to maintaining and enhancing inland fish production and small-scale aquaculture and their associated social and economic benefits are increasing competition for resources and insufficient institutional and political recognition.

Reporting and Statistical Issues in Fisheries

Without reliable statistics, effective fisheries management and policy-making are impossible. During the last decade, financial support for the development and maintenance of national fishery statistical systems have decreased sharply in real terms, while statistical requirements have been increasing dramatically for by-catch and discards, fishing capacity, illegal fishing, vessels authorized to fish in the high seas, economic data (costs, revenues, prices, subsidies), employment, management systems, inventories of stocks and fisheries, aquaculture, etc. Despite FAO's efforts, the available fisheries data are not fully reliable.

The outcome is far from perfect in terms of coverage, timeliness, and quality. Statistics from artisanal and subsistence fisheries are still a concern, and many key statistics are missing, e.g. economic and social data, discards, fishing capacity, etc. The result is that the general trends are probably reliably reflected, but the annual figures and the assessments involve a degree of uncertainty, and small changes from year to year are probably not statistically significant. Illegal, unreported and unregulated fishing is found in all capture fisheries, irrespective of the location, species targeted, fishing gears employed or level and intensity of exploitation. Working with the countries is the only way to improve fishery statistics, primarily to meet national needs with regard to food security and fisheries management. Unlike capture fisheries, the separate monitoring of aquaculture is relatively new in most countries, and often there are less well-established systems of data collection as compared to capture fisheries.

Actual inland capture fisheries production is considerably greater than the amounts reported to FAO. The factor is at least two overall, but may be as high as three in some instances. There is an urgent need for better data on inland fisheries that can be interpreted in both economic and ecological terms. Although the cost of improving inland fishery data collection may be high, failure to fully account for inland capture also is costly in terms of lost opportunities to increase food security and other economic and social benefits from enhanced management of inland fisheries resources.

A recent article in *Nature* indicated that China's marine capture fishery production for 1995-1999 has been overstated in Chinese statistics submitted to and published by FAO. The paper states that as a consequence of this, global marine capture fishery production—excluding Peruvian anchoveta—has likely been declining since 1988 rather than remaining fairly constant as indicated by the statistics. According to the authors, this would have led to understating the degradation of world fisheries, and produced unfortunate policy and investment decisions. However, the FAO maintains that despite likely errors in the data sets, the main global trends have not been masked, and that the most important conclusions have emerged, nevertheless. These findings, together with similar ones emerging at regional and national level, have been the foundation for the governance and institutional changes observed since 1990.

Aquaculture Status and Potential

Aquaculture is more akin to farming and animal husbandry than to fishing, as it involves the rearing and management of living aquatic resources in a restricted environment. Tenure of production facilities, and property rights to the produce, are as important to the success of aquaculture as land tenure is to agriculture. Aquaculture has been developed to serve a variety of purposes: (1) producing high nutritional value food for human consumption; (2) contributing to rural income and employment through farming and related activities; (3) enhancing capture and sport fisheries; (4) cultivating ornamental species for aesthetic purposes; (5) controlling aquatic weeds or pests hazardous to humans or crops; and (6) desalination and other forms of soil recuperation.

Meeting basic human needs for protein foods in the future will be a difficult challenge. Approximately 1.3 billion people live on less than a dollar a day—the cost of a half a pint of beer—and half of the world's population lives on less than 2 dollars a day. Since 1950, there's been a 100% increase in the per capita demand for fish, a 40% increase for grain, and 33% for wood. If world fish consumption will increase from 16 kg (1997) to 19-20 kg by 2030, total human use of aquatic foods will increase to 150-160 million tons. Capture fisheries can provide no more than 100 million tons, so the bulk of the increase will need to come from aquaculture.

In 2001, 48.4 million tons of aquatic products (including plants) valued at US$ 61.5 billion were produced, with half of the production being finfish. Aquaculture is growing more rapidly than all other animal food producing sectors. Worldwide, the sector has increased at an average compounded rate of 9.2% per year since 1970, compared with only 1.4% for capture fisheries and 2.8% for terrestrial farmed meat production systems. Over half of global aquaculture production originated from marine and brackish coastal waters, while the remainder (45%) was from freshwater.

Aquaculture will soon overtake cattle ranching as a global food resource, possibly signaling a basic shift in diets. Over the last century, the world relied heavily on two natural systems—oceanic fisheries and rangelands—to satisfy a growing demand for animal protein, but that era is ending as both systems are reaching their productive limits. Between 1950 and 1990, beef production, four-fifths of it from rangelands, nearly tripled, climbing from 19 million to 53 million tons before leveling off. Since 1990, there has been little growth in either beef production or the oceanic fish catch. Additional production of beef or seafood now depends on placing more cattle in feedlots or more fish in ponds.

Comparisons of energy and production efficiencies of aquaculture versus an array of fisheries and terrestrial agriculture systems confirm that aquaculture is an efficient mass producer of animal proteins for a crowded, coastal planet. Production efficiencies of edible mass for aquaculture range from 2.5 to 4.5 kg dry feed/kg edible mass compared with 3.0 to 17.4 for conventional terrestrial animal production systems. Beef cattle require over 10 kg of feed to add 1 kg of edible weight, whereas catfish can add a kg of edible weight with less than 3 kg of feed. To produce 1 kcal of catfish protein about 34 kcal of fossil fuel energy is required—lobster and shrimp capture fisheries use more than 5 times this amount of energy. Energy costs for even the most intensive salmon cages are less than lobster and shrimp fishing, but are comparable to beef production in feedlots. Aquaculture also has a comparable advantage in water efficiency, since a comparatively smaller amount of water for aquaculture is required. For example, catfish ponds managed using multiple harvest strategies use 1.50 m^3 of water per 1 kg of product while conventional soybean production takes 1.63 m^3 of water to produce 1 kg.

In a world of land and water scarcity, the advantages of aquaculture over capture fisheries and other land-based protein production systems to produce low-cost animal protein are clear. However, care must be taken to ensure that aquaculture is not viewed as a panacea—since most analysts agree that aquaculture will never completely replace capture fisheries—and that increased aquaculture production will not be associated with higher rates of environmental damage and harvest rates of forage fish species used to produce fish meals and oils, which would lead to a net loss of fish production (e.g. the capture for fish meals and oils would exceed the amount of fish produced for consumption).

There are many cases where aquaculture expansion has fueled the hope of fragile coastal and inland rural communities that have undergone unprecedented changes in their traditional ways of life. Aquaculture has provided significant multiplier effects on the local economies increasing both direct and indirect personal spending in these coastal communities. And yet, communities in many parts of the world actively oppose aquacultural development because such development is perceived as a threat to local social and/or ecological systems.

If intensive aquaculture operations—no matter how advanced technically—have no community roots; and feeds, seeds, supplies, equipment and human expertise are procured from great distances from the sites of production; community opposition will continue to occur. In these cases it is easy to see why some community members view aquaculture development as "all we get is your pollution". Planning for aquaculture development as community development and environmental enhancement must thereby encompass regional planning processes to accommodate aquaculture's vital support industries (inputs), and for the use of aquaculture resources and wastes in agriculture or in environmental enhancement projects (outputs). Regional planning for "ecological aquaculture" developments will have much higher positive impacts on jobs and the environment, and will eventually dissolve the opposition from communities who will see the newcomers as one of their own.

Aquaculture production systems, aquatic environments, and the feeds to produce fish must be environmentally benign and ecologically sound in order for the advantages of aquaculture as a food production system to be embraced and supported. Indeed, the expansion of aquaculture can never be justified on the destruction of the world's capture fisheries. The future protein needs of millions of people—and the sustainability of aquaculture itself—depend on the conservation, good management and recovery of the world's capture fisheries and the environments on which they depend.

MARINE BIODIVERSITY

Biodiversity is a concept increasingly seen in scientific publications and on the agendas of environmental protection groups. Scientists have warned that human activities are increasingly cutting further into nature's abilities to respond to external changes by reducing the diversity of living organisms. Loss of biodiversity has reached the agendas of the global community. Initiated by the United Nations Convention on Biological diversity in Rio in 1992, the Conference of Parties of the Convention on biological diversity signed a declaration in Jakarta in 1995 (The Jakarta Mandate). The Mandate sets forth actions needed to preserve biodiversity and emphasizes the need for an integrated management approach for marine and coastal areas. The current status of world fisheries has lent urgency to the goals of ecosystem restoration. Learning how to repair damage and maintain marine biodiversity, while providing access to ecosystem use, has become an important goal of fishery management. Achieving this goal requires understanding the properties of ecosystems, the human uses of ecosystems, and the way those human uses are controlled.

Ecosystem use is controlled through rules and responsibilities of property rights, under the umbrella of the larger institutional setting of national laws and international agreements. Property rights define privileges and responsibilities in the use of environmental goods and services. They specify the way people are to behave toward one another as they use environmental resources (Bromley 1991). Property rights define the conditions that guide and control the human use of the natural environment. They establish the terms under which people use and sustain the capacity of the environment to generate a continuing flow of goods and services. Property rights are a means by which people interact with their environment and control their behavior toward one another. They embody the expectations people have about natural resources and influence the way people make resource use decisions. Property rights are the intersection of marine ecosystems, people and fishery management (Hanna 1996b).

Property rights have several functions. They delineate the population of legitimate owners. They specify the allowable actions of these owners and their associated responsibilities so that expectations are consistent and enforcement of rules is possible. And by setting consistent expectations they reduce uncertainty about others' behavior. In the larger sense, they connect the pieces of the natural system to the pieces of the human system.

The basic functions of managing common pool resources-coordinating users, enforcing rules, and adapting to changing environmental conditions-cannot be met without a system of property rights. The way that property rights function in any particular context determines whether the natural and human system will interact in compatible or conflicting ways.

Property rights in marine ecosystems can control both use and conservation. Their scope may be either individual species or areas of water in which species live. They may also include different types of use. At present, most systems of property rights are designed for direct uses, such as catching fish for food, but they may also include indirect uses, such as the right of the public to enjoy marine birds or to protect endangered species of sea turtles.

The fundamental problem for marine biodiversity protection is that many of the biodiversity values fall outside the realm of direct use, whereas most traditional control rules affect direct use. Direct uses are easiest to monitor and measure, control and trade. Property rights are frequently left undefined for unused species or to communities of species, leaving them unprotected by ownership. Indirect uses such as genetic information and aesthetic appreciation are less visible because their contribution is more diffuse over space and time. Because these resource values are seldom owned, bought or sold, they are often at a competitive disadvantage in a market economy, leaving them undervalued and overused. This leads to the common approach of protecting biodiversity not through property rights but by removing them from human influence in sanctuaries, refuges or reserves.

Are There "Best" Forms of Property Rights for Biodiversity?

Which forms of property rights are best for the protection of biodiversity? People often promote a particular form of property right because of its ability to contribute to achieving larger goals such as community stability, equity or economic efficiency. In fact, the goal of sustaining marine biodiversity can be achieved with a variety of forms of property rights.

The key requirement is that the system of property rights fit into the ecological, economic and social context in a way that contributes to maintaining ecological function and diversity. Property rights take many forms. Differences in form derive from the scope of the rights, the type of rights owner and the privileges and responsibilities of the rights owner. The most common categories are private, public, state and common. The types of property rights are ordered loosely along a spectrum, where the owner ranges from an individual person to no one (Hanna et al. 1996).

Open Access

Open access, *res nullius*, is the absence of property rights and has no ownership assigned. It is open to all, with resources becoming owned only at the point of capture. The absence of rights is often, but mistakenly, referred to as "common property." The "tragedy of the commons" metaphor is a description of the outcome of open access. It describes the use of a common-pool resource for which it is too costly to exclude people from use.

In the tragedy of the commons, individuals make choices about resource use based on their own private costs and benefits. And while the benefits of their actions accrue to them alone, the costs, in terms of the effect on the resource, are social costs spread among all users. Eventually and inevitably without some control over access, the failure to account for social costs locks people into behavior that leads to resource degradation and overuse.

Common Property

Common property, *res communes*, is owned by an identified group of people who have the right to exclude non-owners, the duty to participate in decisions about use, and the responsibility to act as resource stewards. For example, near-shore fishing territories may be owned and managed by the residents of the adjoining coastal community. Community members as owners decide how many people have rights to use the resource, what kind of rights these may be, and what objectives they have for resource productivity. Although the "tragedy of the commons" is a metaphor that is widely used, it does not really describe common property.

State Property

State property, *res publicae*, is owned by citizens who assign management authority and stewardship responsibility to an agency of the state. For example U.S. fish and wildlife resources are the property of the citizens of individual states or of the country as a whole. The citizens designate the fish and wildlife agencies of individual states and the federal government to manage these resources in their name. The agencies may grant rights of use to individual users or communities of users, but the resources remain owned by the citizens at large.

Private Property

Private property, *res privatae*, assigns ownership to named individuals including legal individuals, such as corporations, guaranteeing to those owners a bundle of rights about access and use. Although individuals have the greatest autonomy under private property, this form of property rights is also stinted by prohibitions against unacceptable uses, such as activities that pollute. For example, individuals may own land and the fish and wildlife on that land, but the types of use to which they can put either the land or the wildlife may be constrained by law.

STATUS OF MARINE BIODIVERSITY

The need to maintain healthy ecosystem function is well understood in theory but often poorly achieved in practice. The history of post-World War II fishery management illustrates the problem. Worldwide, fisheries and their management have contributed to marine biodiversity loss and have resulted in large losses in the economic productivity of ecosystems.

The current state of world fisheries is both a cause and an effect of the biodiversity problem. Fisheries are a fundamental cause of the biodiversity problem because of the general focus on single-species management, slow response to environmental signals, and the patterns of sequential harvesting, overcapacity and failure to control access (Hanna 1996). Fisheries bear the effect of biodiversity loss through declining biological and economic productivity, loss of aesthetic and recreational values, a diminished range of harvested species, regulations implemented to protect endangered species, and increasing costs of management designed to protect further biodiversity loss (National Research Council, 1995).

World fisheries developed rapidly after World War II under national policies to promote expanded economic opportunities and control of ocean territory. Access remained open to all, subsidies were provided for the construction of vessels, and seafood markets expanded. As a result, fisheries were transformed from being 60% underexploited in the early 1950s to 60% overexploited by the early 1990s (FAO 1997). FAO reports that of the fish stocks accounting for the majority of the world's marine landings, 60% are in urgent need of management (FAO 1997). The degree of overfishing varies widely by geographic area, but where it exists, species abundance is reduced and ecological resilience is compromised (Botsford et al. 1997). Ecosystem productivity in coastal areas is also suffering from habitat degradation and fragmentation (Gray 1997), and these declines in biological productivity translate directly into economic losses.

Biological and economic stresses on marine ecosystems also affect marine fishery management at all levels. Increasing conflicts, complicated regulations, expanding information requirements and declining legitimacy of management all contribute to increasing fishery management costs. The 1990s have seen a number of publications describing fishery management as in crisis (McGoodwin 1990; Crean and Symes 1996; Hannesson 1996).

Problems in maintaining the biodiversity of marine ecosystems also affects equity and fairness, issues that are especially relevant for food security and employment. Globally, fish are 16% of the animal protein content of human diets. Regionally, fish are a more important protein source for poor countries in Africa (21%) and centrally planned Asian economies (22%) than for rich countries in Western Europe (10%) and North America (7%) (FAO 1993). When biological productivity of marine ecosystems falls, small-scale artisanal fisheries in poor countries suffer loss of subsistence and markets, often without available substitutes (Williams 1996).

MARINE FISH AND CRYOPRESERVATION OF SPERM

Since the first work of Blaxter in 1953, fish sperm cryopreservation has been attempted on about 30 marine species. The present paper reviews the techniques used and the results published in these species. Particular attention is paid to the handling procedure of sperm before freezing, the problems of semen ageing and semen contamination with urine. The quality of frozen-thawed semen was evaluated using previously standardized biotests, such as a two-step motility activation technique adapted for the different species and fertilization assays using a discriminating insemination technique. Most extenders used in marine fish are saline or sugar solutions. From the investigated cryoprotectants, dimethyl sulphoxide (DMSO) generally leads to the best results. Cooling rates range from 8 °C to 99 °C min-1; the thawing rate is generally high. Compared with freshwater species, a high percentage of spermatozoa survives cryopreservation. Therefore, and because of the simplicity of the techniques, the cryopreservation of marine fish sperm is suited for application in aquaculture. Techniques of sperm management have been established in some freshwater fish species such as cyprinids (Billard, Cosson, Crim & Suquet 1995) or siluroids (Legendre, Linhart & Billard 1996) and in salmonids (Scott & Baynes 1980; Billard 1992). Among these techniques, sperm storage and cryopreservation are of special interest. At 0°C conditions, spermatozoa can be stored for a few hours up to several days, depending on the species while cryopreserved gametes can be theoretically stored between 200 and 32 000 years without deleterious effect (Ashwood-Smith 1980).

The use of cryopreserved spermatozoa can be delayed from the date of collection and adjusted to the moment of ova processing. The benefits of this technique include:-Synchronization of gamete availability of both sexes: ovulations are only noticed when sperm production declines in cross fertilization of different strains like spring and autumn spawning herring (Clupea harengus L.: Blaxter 1953).

Use of the total volume of available semen: this is useful for sperm economy in species where semen is difficult to obtain (Japanese eel, Anguilla japonica Temminck & Schlegel) (Ohta & Izawa 1996), but also in species where only low volume of semen can be stripped in captivity (yellowtail flounder, Pleuronectes ferrugineus L.: Clearwater & Crim 1995 or turbot, Psetta maxima L.: Suquet, Billard, Cosson, Dorange, Chauvaud, Mugnier & Fauvel 1994).-Simplifying broodstock maintenance: off season spawning can be induced in most cultured fish species, by the manipulation of photoperiod and temperature cycles (Bromage 1995). However, the technique is cost intensive. When cryopreserved sperm is available all year round, the manipulation of the spawning season could be restricted to females.-Transport of gametes: useful when male and female gametes are collected in different locations. This enables also the introduction of genes from the wild into hatchery stocks.

Avoiding aging of sperm: the senescence of sperm during the course of the spawning season has been reported for many fish species and results in a decrease of milt quality (Rana 1995a).

Cryopreservation allows the collection of sperm when it has the highest quality.

Experimental programs: for genetic studies, in comparing the breeding performances of successive generations in the same experiment and for experiments where the use of identical sperm samples is necessary over an extended period e.g study of short term storage of ova. Conserving genetic variability in domesticated populations: the use of a limited number of breeders leads to a reduction of heterozygosity. The cryopreserved semen of selected strains or genetically improved populations can be introduced in domesticated stocks e.g. the sperm of sex reversed gynogenetic hirame (Paralichthys olivaceus Temminck & Schlegel) female (Tabata & Mizuta 1997). Gene banks of cryopreserved semen can also be used to maintain genetic diversity of fish populations that are endangered and protect against inbreeding. In protogynous hermaphrodite species such as Black grouper (Epinephelus malabaricus Bloch & Schneider), sperm can only be collected in 5 to 10 years old animals (Gwo 1993). As a consequence, success in breeding is greatly enhanced by the use of frozen sperm.

Under consideration of the many benefits of this technique, the sperm of over 200 fish species has been cryopreserved . These techniques have been reviewed in detail by Stein & Bayrle (1978), Scott & Baynes (1980), Stoss (1983), Chao (1991), Jamieson (1991), Tiersch (1995), Rana (1995a & b), Maisse (1996) and Maisse, Labbé, Ogier de Baulny, Leveroni & Haffray (1998). Most publications are devoted to salmonids, tilapias and carps. In marine fish, the first report was by Blaxter (1953) in herring. As marine fish farming expands, there is an increasing need to apply sperm cryopreservation techniques. Due to the increasing knowledge in sperm biology also spermatozoa of other marine fish have been recently cryopreserved successfully. This paper reviews the techniques and results obtained for cryopreservation of marine fish species. It emphasizes the high survival and fertilization capacity generally obtained in frozen-thawed spermatozoa of marine fish in comparison to freshwater species. Special attention was paid on individual variations of sperm quality and on biotests useful to determine the quality of thawed cells.

INDIVIDUAL VARIATIONS OF SPERM QUALITY

High individual variations of milt quality are frequently reported (Rana 1995a). This may be due to genetic variability but also to sampling location (from testes to the genital pore), contamination of sperm samples by urine and aging of spermatozoa during the spawning season. Similar to the Salmonidae (Morisawa & Morisawa 1988), a maturation process of spermatozoa has been reported for several marine fish species. Compared to samples collected by stripping, the percentage of motile spermatozoa of turbot decreases when intratesticular milt is collected. In Japanese eel, testicular spermatozoa exhibited only a very slight motility after activation, while spermatozoa of stripped sperm could be successfully activated (Ohta, Ikeda & Izawa 1997a). Furthermore, testicular spermatozoa of this species, incubated in saline solutions with high concentrations of K_+ and HCO_3^-, artificially acquire their motility potential and show a similar fertilization capacity as stripped milt (Ohta, Tanaka, Kagawa, Okuzawa & Iinuma 1997b). Because of the close vicinity of sperm ducts and urinary ducts, sperm samples are frequently contaminated with urine. This phenomenon is described in freshwater fish species (Rana 1995a), but more scarcely in marine species. In turbot, a mean contamination rate of 15.3% (urine volume: sperm volume) was reported, basing on the determination of urea concentration (Dreanno, Suquet, Desbruyères, Cosson, Le Delliou & Billard 1998). Artificial urine contamination of milt decreased the percentage of motile spermatozoa, sperm velocity, the fertilization capacity and the storage ability. These effects increased with urine concentration and incubation time. Catherization of the ureter prior to sperm collection significantly decreased the urine contamination to 9.3%. In species with high sperm densities such as sea bass (Dicentrarchus labrax L.), urine contamination is more easily detected because of the lower viscosity and colour changes of portions of sperm samples containing urine.

Intratesticular aging of sperm has been reported for many fish species and it affected sperm quality at the end of the milting period (Rana 1995a). In marine fish, in the sea bass the concentration of spermatozoa decreased as the spawning season progressed (Fauvel et al. 1999). Spermatozoa of this species maintained their swimming duration for a longer period at the beginning of the milting season than at the end. Furthermore, lower motility rates, fertilization rates and reduced short-term storage capacity were recorded at the end of the reproduction period. As revealed by electron microscopy, also the structure was changed in aged sea bass sperm (Dreanno, Suquet, Fauvel, Le Coz, Dorange, Quemener & Billard in press).

In the Atlantic halibut (Hippoglossus hippoglossus L.), sperm motility was reduced at the end of the reproductive period (Methven & Crim 1991; Shangguan 1998). The highest motility rates of winter flounder (Pleuronectes americanus Walbaum) sperm were found at the beginning of the milting period (Shangguan & Crim, 1995). A decline of motility parameters, storage and fertilization capacities as well as alterations of the plasma membrane were reported in turbot as the milting season progressed (Suquet, Dreanno, Dorange, Normant, Quemener, Gaignon & Billard 1998a). As recorded in sea bass and turbot, senescence of spermatozoa can result in a decrease of freezing ability. Subsequently, when sampling sperm for cryopreservation, one must take care of the three described factors (intra testicular maturation, contamination by urine and aging of sperm) as they can decrease the initial quality of spermatozoa.

BIOTESTS

Several biological tests are used to improve the cryopreservation techniques and to assess the sperm viability of frozen-thawed spermatozoa. An efficient quality test for frozen-thawed spermatozoa must be correlated with the fertilization capacity. The test must allow objective and sensitive measurements of a large number of cells and give a rapid answer because of the fast deterioration of ova quality immediately after collection. The presently applied tests describe physiological parameters (percentage of motile spermatozoa and fertilization capacity). Complementary biochemical tests (ATP, O_2 consumption) or morphological ones (electron microscopy) precisely localize damages due to the cryopreservation process.

Motility Assessment

Billard & Cosson (1992) reviewed and improved early techniques used to describe fish sperm motility: because of a high sperm concentration in milt, the assessment of sperm motility requires a high dilution rate. A two-step procedure is necessary to initiate simultaneous motility of the maximum number of spermatozoa: sperm is first diluted in a medium that does not initiate motility. Then, sperm movement is activated, by mixing spermatozoa directly under the microscope with an activating solution.

Cosson, Billard, Cibert, Dreanno, Linhart & Suquet (1997) developped the techniques allowing a fine description of motility. This observation is greatly enhanced using dark field optics microscopy attached to a camcorder. Spermatozoa can be visualised on a video monitor. Using stroboscopic illumination, the flagellar beat frequency can be measured by reference to the calibrated frequency of the flash illuminator. The percentage of motile cells is estimated using an arbitrary scale, each step representing a wide range of spermatozoa showing progressive movement. CASA (computer-assisted sperm analysis) system allows an analysis of videotapes using software which set up parameters are adjusted for the studied species sperm characteristics. Straight or curvilinear velocities are rarely assessed using CASA system (turbot, Dreanno, Suquet, Quemener, Cosson, Fierville, Normant & Billard 1997; sea bass, Fauvel, Suquet, Dreanno, Zonno & Menu 1998a). Among the criteria used to estimate sperm motility, the percentage of motile cells is significantly correlated with the fertilization capacity of sea bass and turbot spermatozoa.

Because no significant correlation was reported between fertilization rate and the velocity of spermatozoa in these species, the success of fertilization was assumed to depend on the number of motile cells rather than on the quality of sperm movement (Dreanno 1998). However in cod (Gadus morhua L.), the proportion of spermatozoa exhibiting a progressive movement accounted for only a small percentage of the fertilization success (Trippel & Neilson 1992), and spermatozoa showing non progressive vibrating motion could fertilize ova of this species. As a consequence, complementary tests must be used to assess the quality of frozen-thawed spermatozoa.

Assessment of energetic status: the determination of intracellular ATP concentration has proven to be informative in freshwater fish species, because a drastic decrease of nucleotide content has been reported in frozen-thawed spermatozoa compared to fresh sperm (Ogier de Baulny, Le Vern, Kerboeuf, Heydorff & Maisse 1996). Furthermore, a significant positive correlation was reported between the ATP content of turbot spermatzoa and the percentage of motile cells. In sea bass, Adenylate Energy Charge (AEC) is correlated with the percentage of motile cells and the fertilization capacity (Dreanno 1998). The measurement of intracellular ATP content at various times post-activation and the assessment of oxygen consumption allow to assess the degree of cell integrity before and after the cryopreservation (Dreanno et al. 1997).

Seminal fluid composition: in rainbow trout (Oncorhynchus mykiss) seminal plasma pH and osmolality have been selected to determine the semen suitability for cryopreservation (Malejac, Loir & Maisse 1990; Lahnsteiner, Berger, Weismann & Patzner 1996a). A high content of 42kd protein in the seminal fluid was also correlated with a low fertilization capacity of frozen-thawed rainbow trout spermatozoa (Maisse, Pinson & Loir 1988). The oxaloacetic transaminase activity of the seminal fluid of thawed rainbow trout spermatozoa was also considered as a quality test (Malejac et al. 1990). The leakage of these different components into the seminal fluid reflects the alteration of the sperm membrane during the cryopreservation.

***Swelling tests*:** This technique derives from rainbow trout. Sperm samples were incubated in hypo-osmotic solutions and their resistance to the osmotic shock correlated with the cryopreservation ability (Malejac et al. 1990). However, this test is difficult to use in turbot, because no lysis of spermatozoa was observed after dilution in freshwater (Dreanno 1998).-Fertilization capacity: because of the difficulty to collect good quality ova, the fertilization capacity of frozen-thawed spermatozoa has only been tested in a few studies.

However, it represents a major step forward in the management of gametes. A standardized experimental insemination procedure is necessary for reproducable evaluation of the fertilization capacity (Suquet, Gaignon, Quemener & Normant 1995a).

The problem of replicate variation is illustrated by the sampling technique of ova. When a standard pipette was used, more ovarian fluid than ova was sucked in, resulting in a large methodological error (coefficient of variation: 66.5% in nine successive 800μl samples) in the number of sampled ova. This variation was largely reduced using a positive displacement pipette (5.1%). Weighing ova samples may also be useful. However, correlations between the number of ova and their weight must be calculated for each species and variations could also be due to individual fish conditions or seasonal changes.

The fertilization capacity of frozen-thawed spermatozoa has to be tested using discriminating conditions. In rainbow trout, a significant correlation was reported between fertilization rate and motility estimates using a low number of spermatozoa per ovum (<200,000), whereas a supraoptimal ratio did not allow to detect this correlation (Moccia & Munkittrick 1987). Therefore, knowledge of the optimal sperm to ovum ratio is necessary to improve the cryopreservation technique. Optimal sperm to ovum ratios are only reported for few marine fish species in the literature.

CLIMATE, BIODIVERSITY AND ENVIRONMENTAL CRISIS

Climate change is more than evident in many parts of the world, characterized by altered patterns of natural phenomena, severe damages and human casualties (IPCC, 2007). Moreover, ice melting and the resulting rise in sea level increasingly threat coastal lowlands due to flooding, erosion and salinization, and further alteration in the balance of aquatic systems (Handyside *et al.*, 2007). Temperature rise will particularly affect freshwater fish (Co-chrane *et al.*, 2009) and the problem is expected to be severe, for example, for the 37% of the species in Greece that are already classified as threatened. In marine and brackish waters, tropical and subtropical alien species have extended their distribution pattern to the north of the Mediterranean (Zenetos *et al.*, 2009). Overall, climate change is expected to favor many of the 11000 alien species recorded in Europe (EU, 2008).

North American non-indigenous crayfish species (NICS), which are asymptomatic carriers of the crayfish 'plague', were imported to Europe (via aquaculture, aquarium trade, as live food, and for re-stocking purposes) devastating in many countries the indigenous crayfish species (ICS) due to their aggressive behavior, higher growth rates and fecundity (Holdich *et al.*, 2009). It was projected that European waters will be inhabited by NICS in 100 years time and few, critically endangered, ICS populations will be confined in limited isolated places (Taugbøl and Skurdal, 1999).

Although the 'polluter pays' principal has been welcomed by environmentalists, ecological footprint data revealed that the problem was particularly urgent in resource use. In Greece, the overall ecological footprint is about 3.5 times higher compared to the national available biocapacity, and about 2.2 times higher compared to the global ecological footprint (WWF, 2008). Footprint in aquaculture ranges from negligible to as much as 50000ha per ha activity, depending on species, farming methods and intensity (Beveridge *et al.*, 1997; Folke *et al.*, 1998). In fact, the ratio of wild fish inputs to farmed fish outputs in the intensively farmed carnivorous species (e.g. Atlantic salmon) is 5.0 compared to the average value of 0.63. Fish oil imports are dominated by Europe and volume requirements by the aquaculture sector is expected to determine the sectors role in the future of wild fisheries (Naylor *et al.*, 2009).

NUTRITIONAL CRISIS AND FISCAL GROWTH: INDIA ON THE EDGE

Annual aquaculture growth in India follows the national economy trend (~7% during 2007-2009; EW, 2010; FAO, 2010). In the same time, 36% of the population lives below poverty line and 48% of the children of 0-5 years old are chronically malnourished (Arnold *et al.*, 2009). Trade liberalization generated revenues to the national economy from shrimp exports, but at the same time it was responsible for an imbalanced growth and poverty increase at local level (Pradhan and Flahenly, 2008). In the agriculture sector, farming of genetically modified organisms (GMOs) by small farmers has compromised their livelihoods, since they cannot afford to pay for GMO grains, inorganic fertilizers, and pesticides. The overall picture implies that nutritional deficiencies and fiscal development can go hand to hand, especially when agricultural products became the new battleground for agro-fuel companies and stock exchange speculators.

Social Crisis

The collapses of Bear Sterns and Lehman Brothers opened the Pandora's Box for the globalized economic networks. Whether this is a decaying over-accumulation (capital concentration) crisis (Marx and Fowkes, 1992), followed by a reproduction crisis of the new high-tech era, or just a creative destruction phase of the previous economic cycle (Schumpeter, 1962), the current crisis has affected the real economy and society. Total unemployment rate reached 17.5% in the USA (Leonhardt, 2009) and 13.2% of the population (40 million people) live in poverty (DeNavas-Walt *et al.*, 2009). This trend is expected to be accompanied by a parallel under-consumption crisis, and negatively affect consumption habits in fishery products. Currently, the predicted needs for fish food supplies are huge (i.e. 270.9×10^6 metric tons for 2050; Wijkström, 2003) and the aquaculture sector is expected to fill the demand gap. However, even if it is possible to overcome the finite character of natural resources, little evidence exists at present that *per capita* supply and consumption of fish products will rise, for example, in the sub-Saharan countries (~less than 8kg *vs* 16.5kg globally), suggesting that food security and accessibility opportunities even to basic goods are not related to the monetary gain-driven nature of the industrial aquaculture and fisheries sectors.

Fisheries in the Twilight Zone

Overexploitation of commercial fish stocks in most of the seas is mainly the result of industrial trawling (Watling and Norse, 1998). The collapse of Northeast Atlantic cod fishery and processing industry in New England during the 1990s (Hamilton *et al.*, 2004a) was not a sudden event. Early signs of stress in Atlantic cod, halibut and other commercial stocks were evident from 1930s (Clausen and Clark, 2005). The evolution of fishery practices and technology resulted in mass production and decimation of large predators and, subsequently, to extracting fish of the lower trophic levels. It is now widely accepted that overfishing of predators leads to disruption of the energy flows between the food webs, increased by-catch and loss of biodiversity, and, finally, to collapse of the entire vulnerable ecosystems.

The ecological consequences of both 'fishing down the food chain' (Pauly *et al.*, 1998) and 'fishing down the deep' (Morato *et al.*, 2006) practices will have ultimately serious effects on populations which rely on fisheries. During the

1990s, overfishing and environmental stress in the North Sea caused depletion of commercial stocks and, accordingly, generated increased unemployment, business failures, and rapid structural demographic disturbances in the Faroe Islands (Hamilton *et al.*, 2004b). Currently, in developing countries that collectively represent 70% of total catch and 35% of small scale fisher-ies, 22-24 million fishermen and 68-70 million people who work in the postharvest sector are expected to face serious consequences from the fish supply crisis (Hall *et al.*, 2010). Political will at the international level currently refects the dominance of fish-ery companies and trade laws over environmental laws, and the future of marine capture fisheries will be ultimately conditioned by political, social and economic factors, including aquaculture (Garcia and Grainger, 2005).

Metabolic Rifts in Industrial Fisheries and Aquaculture

Sustainable development principles have become the mainstream doctrine. However, the dominant neoliberal attitude and practice towards the environment (i.e. continued economic expansion in order to built capital for investment) has produced irreversible environmental disasters.

Material exchange in biological chemistry and agriculture, where depletion of soil nutrients was observed early (mid 18th century), led Carl Marx to deploy the concept of metabolic rift. Capitalism is the driving force for centuries and organizes social metabolism with nature on the base of capital, production and proft escalation. However, the equilibrium of these dynamic metabolic processes between humans and nature are frequently ruptured, as constant inputs of fnite resources are required and wastes are accumulated. Such a rupture (metabolic rift) also threatens the stability and biological integrity of aquatic ecosystems (Clausen and Clark, 2005). Accordingly, industrial aquaculture ('Blue Revolution') is challenged as a 'quick-fix' to transcend ecological limits and existing metabolic rifts such as

a) The development of shrimp farming by the construction of mega-farms both in Central America and Asia. In India, Bangladesh, Thailand and many Southeastern Asia countries, for example, shrimp farms were built on unfit acidic soils after clearing mangrove forests (Barbier and Sathirathai, 2004). Moreover, post larvae production is entirely based on wild brood-stock (Primavera, 2006). The use of open farming systems resulted in continuous degrading water quality (Naylor *et al.*, 1998). Finally, the shrimp sector in most countries has been devastated by the infection by White Spot Syndrome Virus (Flegel and Alday-Sanz, 2007), with serious soioeco-nomic impacts to local communities and economies.
b) The unprecedented exploitation of glass eel stocks in European waters was evident during the last 15 years, in order to fulfll the demand by

Chinese farms; it resulted in sky-high prices glass eels (>€1000/kg) and dramatic decrease of natural stocks in European inland waters (ICES, 2008). Accordingly, eel was registered in the IUCN list of threatened species.

c) The exploitation of mature stocks of Mediterranean blue-fin tuna, which have almost crashed and most specimens caught are far below the reproductive age (MacKenzie *et al.*, 2009). Tuna farming for sushi depends solely on state quotas in Europe, as well as on fish caught in the Mediterranean waters of North Africa. Recently (on March 2010), the involved parties (EU, USA, Japan, Canada, etc.) failed to halt tuna fishing during the last CITES meeting, by-passing the urgent alerts of most fishery scientists who suggested zero quotas (Black, 2010), while favoring the fishing industry and their lobbies.

SOCIOECONOMIC STATUS OF MARINE FISHERMEN AND THEIR UPLIFTMENT

Bangladesh is endowed with vast inland and marine water resources where large number of fishermen are engaged in fishing. Fisheries sector provides full time employment to an estimated 20 million people comprising of full time fishermen, small fish traders, fish transporters, processors, packers etc, (World Bank 1989). Beside this, about 10 millions people are directly or indirectly involved in fishing and/or related occupation (Nuruzzaman 1990). The number of fishermen in different sub-sector of fisheries increased rapidly for the last few years.

In the Fourth Five Year Plan (FFYP), the Government of Bangladesh set priority to increase the fish production from 847000 mt to 1200000 mt within the period 1989–90 to 1994–95. The main objective of FFYP is to improve the socioeconomic condition of fishermen and others engaged in fisheries and create additional employment opportunities for poverty alleviation. In such situation, socioeconomic conditions of fishermen is needed to be overviewed to know how they are and what steps should be taken to improve their socioeconomic status. The main purpose of this present paper is to examine the socioeconomic status of marine fishermen where large number of fishermen are involved for their livelihood and to make policy recommendation for their upliftment.

The marine fisheries comprises with two subsectors-artisanal fisheries and industrial fisheries. Small-scale artisanal fishing is carried out mostly by traditional mechanized boats and industrial fishing is done by trawler. There are 52 trawler and 3317 mechanized boats which are used in marine fishing in Bangladesh. The number of motorized boats used for small-scale marine fishing are same for the period 1985–86 to 1988–89 but from other sources (World

Bank 1991, Chong et al 1992, Islam and Elahi, 1993) indicate that the actual number of boat is more than 5000 and most of them are unregistered. 1.3 million fishermen were engaged in fishing, of which 60% and 40% in inland and marine fisheries respectively. In inland capture fisheries, number of fishermen increased by 6% from 1979–80 to 1988–89 but in the same period, the marine fishermen increased by 50% which indicates that more fishermen are getting engaged in small-scale marine fishing. In Bangladesh, marine fisheries is dominated by small-scale marine fisheries. It contributes about 96% of the total marine catch and provides employment to 497000 full-time and 2525000 parttime fishermen (Chong et al, 1991).

Artisanal Marine Fisheries and Fishermen

Traditionally, all kinds of fishing were practiced at subsistence level by the 'jaladas' of the Hindu community. But with the increasing demand and commercialization of inland and marine fisheries and decrease in land assets due to population pressure, a large number of Muslim fishermen took up fishing as a full-time job. Most of the small-scale fishing boat owners either motorized or non-motorized, are now Muslim who hire Hindu and Muslim fishermen as crew, mostly on a catch share basis. Generally, there are two categories of fishermen in smallscale marine fishing, they are boat owners and crews. Most of the boat owners do not go to sea, they are non-sea going fishermen. But they provide mechanized boat, gear and other necessary fishing materials and the crews work as actual fishermen and undertake different responsibilities in fishing.

The income sharing arrangement among boat owner and crews depends on the type of gear used and period in fishing season. For instance, in Chittagong and Cox's Bazar, 60% of the value of catch, after covering the operation costs, goes to boat owner and the remaining 40% is distributed among the hired crew and the share of each crew member depend on the type of functions performed by him (Rahman, 1993,). These may include catching, drying, transporting and support services such as cooking. Very few surveys have been conducted on small-scale marine fishing. Thomson, Jahan and Hossain (1993) made socioeconomic survey on estuarine set bag net fishermen. Islam and Elahi (1993) studied the socioeconomic condition of different categories of small-scale marine fishermen involved in gill net, long line and set bag net. Fishing team, usually consists of one head mazhi, one assistant head mazhi, one driver and 12 to 15 unspecialized fishing labourers. Depending on population involved in fishing team, 111 head mazhi, 26 assistant head mazhi, 69 driver and 544 fishing labour were selected randomly and interviewed to know their socioeconomic condition.

DETERMINANTS OF SOCIOECONOMIC STATUS OF MARINE FISHERMEN

Family Size and Composition

The family size and its composition are related to occupation and income. The study observed that family size was small with low paid jobs. Boat owners whose incomes were highest have the largest family size. The average family size was 8.14 persons of whom 56% were male and 44% were female. The labourers had the smallest family size. They had average family size of 6.37 of which 58% were male and 42% were female. The average family size appears to be higher than the national average. It should be noted that the percentage of female members is lower than that of male members in all categories of fishermen.

Employment and Occupational Structure

Annual fishing activities are divided into three sub-periods. The high activity period (HAP) includes the months of September, October and November, the medium (MAP) includes the months of December, January, February and March, and the low (LAP) includes five months from April to August. The low activity period, June and July, (rainy season), is the slack period for fishing. However, there is little variation of employment during the activity periods among different categories of fishermen. Head majhi, assistant head majhi and labourers spent obove 90% time on fishing.

The boat owners and other fishermen were relatively less involved with fishing. Boat owner's second most important job was business while other fishermen were involved with different occupations is varying period. Thomson et al, (1993) indicated that 35% of the fishermen population were involved in productive activities, 37% were children, below the age of 10, and 28% did not have any employment. There was high rate of dependants on the income earning members.

Annual Income of Different Categories of Fishermen

It is seen that different categories of fishermen earned about 90% of their incomes from fishing. Other sources viz, agriculture or business did not contribute much. Assistant head majhi earned more than 99% income from fishing. Business was the second important source of income in cases of boat owners, head majhis and other fishermen. The level of annual incomes and their differences among different categories of fishermen are considerable. The annual income of boat owner is about Tk 461304. But the fishermen get Tk. 13568 annually. The average annual per capita income of different categories of fishermen household also varies significantly.

Expenditure Pattern and Surplus Income of Fishermen

Monthly per capita expenditure in different categories of households. The expenditures were divided into five categories of which four relate to basic human needs-food, clothing, education and medication. The boat owners used more than 56% expenditure on non-basic items while fishermen used most of their income on the basic items. Food is the single most important category of expenditure. The information confirms the Engel Law that expenditure on food varies inversaly with the income. While boat owners spend 34% on food, whereas fishermen use more than 82%.

Assets

It is reported that the boat owners had household assets valued at Tk. 24 lakh per family or more than Tk. 5 lakh per capita. On the other hand, other fishermen had household assets valued at Tk. 22000 per family or more than Tk. 6000 per capita. However, most of the marine fishermen have no fisheries assets. Boat owners provide boat, gear and other necessary assets for fishing. But Thomson et al (1993) reported that 82% of estuarine set bag net fisheemen are owners of their gear and craft.

Another survey which was conducted in 1985 (Huq et al, 1986) in Cox's Bazar areas, reported that, almost all fishermen had homestead area about 20 decimals or less. About 97% of households had homesteads upto 10 decimals only. As for the agricultural land endowments, 92% of households were landless. However, substantial percentage of fishermen households did not own any boats or nets.

Problems and Constraints of Small-Scale Marine Fishing

a. It was reported that price of fuel and net material have increased rapidly during few years, which seriously affected the net income of fishermen.
b. Quality timber, resulting in difficulties in building new boats and repairing existing ones.
c. Frequent engine trouble interrupts fishing.
d. Lack of capital to construct a boat and to buy engine, gear and other necessary materials. To get credit from the Banks is very complicated and normally not available in time.

Other major problems are piracy, bad weather, lack of timely weather bulletine and no basic medical or insurance policy for fishermen.

CONTAMINANT IN SEAFOODS

Among organic compounds, polychlorinated biphenyls, dioxins, several chlorinated hydrocarbon insecticides, certain processing-related contaminants (nitrosamines and possibly products of chlorination), and contaminants related to aquaculture pose sufficient potential risks for consumers to be worthy of additional study. In addition to providing a broad survey of data on chemical contamination of aquatic organisms and potential risks, this chapter undertakes an extensive set of analyses of the variability of concentrations of certain contaminants across geographic areas and the implications of this variability for control. In general, lognormal distributions appear to provide good descriptions of the pattern of variation of chemical contaminant concentrations among different geographic areas, and some contaminants (mostly organics) appear to be much more variable than others. The variability of contaminant concentrations among geographic areas is important because it indicates the potential for reduction of exposure through restrictions on the harvesting of aquatic organisms from specific sites. Based on analyses of data for inshore marine waters, for the most variable contaminants/sets of species, it would be possible theoretically to reduce the population dosage delivered by more than 50% by restricting harvesting/marketing from only the 5% most intensely contaminated sites.

There is, therefore, considerable potential for management of the overall population dosage of contaminants by measures that would restrict harvesting in specific ways. There is no area of the committee's work that poses greater challenges to both the scientific tools for understanding likely health hazards and the social tools for managing risks, than the diverse collection of chemical residues that find their way into the human diet partly by way of seafood. Moreover, the confusion between technical questions and social control problems is connected. The understanding of toxicology and environmental health has made important strides since the multitiered structure of federal food protection law was erected (principally by legislation in 1906, 1938, and 1958). Older concepts, which shaped the legislative framework within which food protection agencies attempt to function, suggested sharp distinctions between "poisons" and other substances, or between "safe" and "unsafe" levels of exposure to important categories of environmental toxicants. These ideas are gradually giving way to a more quantitative (although generally still highly uncertain) conception of risks, based on more detailed information about the mechanisms by which different substances interact with intricate biological systems and the diversity of those systems in different individuals in the large and disparate human population. To the extent that increased understanding indicates that certain categories of risks cannot be eliminated entirely, the tools for social control of these risks will have to be adapted to manage toxicant exposures and risks in the light of explicitly formulated trade-offs between the costs of forgoing certain portions of our food resources and the costs of potential adverse effects.

The technical advances that have occurred in risk assessment in recent years have been applied most readily to issues of health protection by governmental institutions of relatively late vintage, operating under legislation adopted within the last 20 years – most notably, the different branches of the U.S. Environmental Protection Agency (EPA) and analogous state authorities. When the more modern techniques and assumptions for quantitatively assessing risks are applied to seafood contaminants, there are a number of areas of mismatch that give the appearance of inconsistency in the social and technical judgments on risks made by different agencies. Both this chapter and the next deal with aspects of chemical residues in aquatic organisms. In this chapter the committee focuses on the tasks that are usually thought of as part of the hazard identification portion of a quantitative chemical health risk assessment. The committee has not, however, attempted a formal and comprehensive assessment of the risks of chemical residues in aquatic organisms. Aside from the fact that the available data on both contaminant levels and risks are inadequate for such a task at this time, the charge to the committee emphasized the review of the adequacy of current risk recognition, risk assessment, and risk management procedures in governmental agencies. In the next section of this chapter, the committee gives a broad overview of the types of toxic agents that are known or believed to be contaminants of seafood. Then, the various data bases available for characterizing the geographic and species distribution of chemical contaminants are reviewed, followed by the quantitative insights gleaned from these sources. The variability of contaminants by geography and species, which provides some of the most potentially important opportunities for reduction of exposures, is then considered. Finally, preliminary conclusions and recommendations are offered, based on the material discussed in this chapter.

Metals and Other Inorganics

The human and veterinary medical literature is replete with information regarding the toxicity of heavy metals. Based on this information, different metals can be classified as having major, modest, minor, or no potential for toxicity. Those with major potential for toxicity, in the committee's view, are antimony, arsenic, cadmium, chromium, lead, mercury, and nickel. Contaminants with a modest potential for toxicity include copper, iron, manganese, selenium, and zinc. Those of minor or no toxicity are aluminum, silver, strontium, thallium, and tin. This classification is based, among other parameters, on potency for producing effects and accessibility of the toxicant. Thus, such metals as nickel and chromium, known inhalant carcinogens, are among those of greatest toxicity, whereas selenium and tin are placed in the lesser categories. When considering the same metals as contaminants of an aquatic food source, however, their relative toxicities will certainly change. Criteria for identifying contaminants (hazard assessment, hazard analysis) of public health concern in the aquatic environment may vary but have been defined (PTI, 1987). These include (1) persistence, (2) bioconcentration potential, (3) toxicity to humans (or suspected toxicity), (4) sources of contaminants in the area of interest, and (5) high concentration in fish and shellfish from the area of interest. By applying such criteria, both nickel (except for its carbonyl form) and chromium (at least in its hexavalent form), inhalant carcinogens and elicitors of dermal hypersensitivity, would be suspect as contaminants of public health concern. However, both are poorly absorbed from the gastrointestinal tract, and there is little evidence that this route of exposure results in systemic toxicity. Similarly, the use of tri*n*-butyltin (TBT) to control marine fouling of vessels and aquaculture sea pens has been followed by the accumulation of butyl- and elemental tins in the muscles of fish and invertebrates.

Organic tin compounds tend to be more toxic than inorganic salts, and organic forms in particular may be of public health concern. Although little information exists about the toxicity of tin to man, there is sufficient information regarding dosage levels without observable effect to eliminate the probability of tin poisoning from contaminated seafood. Conversely, selenium is well recognized as toxic by ingestion and, at existing levels in some seafoods, may be a source of risk. Antimony has been recognized as both an occupational and an iatrogenic toxicant. Recent seafood residue studies, however, either have failed to sample for this metal or indicate concentrations above detectable levels in few contaminated sites. Such findings and reasoning, coupled with estimates of ingestion levels, suggest a preliminary list of heavy-metal contaminants found in the edible portions of aquatic animals that may be detrimental to human health. The metals identified in this hazard analysis include arsenic, cadmium, lead, mercury, and selenium. Some toxicity information related to antimony is given later, in the section where current dosage is compared to "acceptable daily intake" levels and other recommended standards.

SPECIFIC TRACE METALS

Arsenic

Arsenic has a long history as a potent poison of humans and other animals. Previously used as a chemotherapeutic and homicidal drug, much information has been collected regarding its toxicity. It exists as the toxic trivalent form (arsenic trioxide, sodium arsenate, arsenic trichloride, etc.), as the less toxic pentavalent form (arsenic pentoxide, arsenic acid, lead arsenate, calcium arsenate, etc.), and as numerous organic forms (arsanilic acid, bimethyl arsenate, etc.). When ingested, inorganic arsenic may cause acute or chronic toxicity and is of primary concern as a carcinogen responsible for pulmonary carcinoma, hemangiosarcomas, and dermal basal cell and squamous cell carcinomas. Its toxicity is dependent on oxidation state and route of exposure. In its chronic manifestations, arsenic is responsible for gastroenteritis, nephritis, hepatomegaly, peripheral symmetrical neuropathy, and a number of lesions of the skin including plantar and palmar hyperkeratosis and generalized melanosis.

Some of these lesions appear related to destruction of capillary endothelium, with consequent edema and even circulatory failure. At the molecular level the metal is known to uncouple phosphorylation; to react with sulfhydryl groups, thus upsetting cellular metabolism; to damage deoxyribonucleic acid (DNA) directly and to inhibit its repair (Buck, 1978). In addition, as sodium arsenate and arsenite it is teratogenic in lower animals (Earl and Vish, 1978). The metal, therefore, places at special risk pregnant and nursing mothers and their children. However, the predominant form of arsenic that exists in the edible portions of aquatic animals is the organic form, either arsenobetaine or arsenocholine. These forms have been named "fish arsenic" and no toxic effects from their ingestion have been reported in animals [at doses of 10,000 milligrams (mg) per kilogram (kg)] or in humans. Furthermore, there is no evidence of mutagenicity by arsenobetaine (Penrose, 1975; Tam et al., 1982). Although arsenobetaine constitutes the bulk of arsenic in fish, available studies are inadequate to conclude that the amounts of more toxic inorganic forms of arsenic (or organic forms that can be metabolized to inorganic arsenic in humans) are negligible in all fish. It is known, however, that the trivalent form (inorganic) is toxic to man and that long-term effects include dermal hyperkeratosis, dermal melanosis and carcinoma, hepatomegaly, peripheral neuropathy, and in cases of inhalation, pulmonary carcinoma. Arsenic is used in the manufacture of pesticides, herbicides, and other agricultural products and is a by-product of mining and smelting operations.

Cadmium

Cadmium is unique among toxic metals because it is a relatively recent (50 years) contaminant of the aquatic environment. Its sources are solid waste dumping (pigment in paint) and cadmium-containing sewage sludge, the use of phosphatic fertilizers, electroplating and galvanizing manufacture, and mining (zinc, lead) wastewater. Cadmium is commonly found in its metallic form and as sulfides and sulfates. Invertebrates, both crustacea and bivalves, tend to accumulate metallic cadmium in large amounts by binding to various high-molecular-weight metallothioneine ligands. There is a differential affinity between crustacean muscle and hepatopancreas, the latter organ containing

10-20 times the concentration of the former. Because hepatopancreas may be considered a delicacy or marketed as "brown crab meat," the potential for ingesting large amounts of cadmium when eating lobsters or crabs is increased.

Cadmium may damage cells by its activity in the plasmalemma where it reacts with phosphate groups of the lipid bilayer to alter permeability, in the nucleus where it is mutagenic, on lysosomal membranes, and as an inhibitor of mitochondrial activity. Its ability to stimulate metallothioneine production in aquatic animals, however, does much to decrease its toxicity (Viarengo, 1985).

Cadmium has been responsible for major human poisoning incidents as a contaminant of wastewater used for irrigation in Japan where the illness is known as itai-itai (ouch-ouch) disease. It is a chronic osteoporotic and osteomalacic condition that primarily affects multiparous females (Kobayashi, 1978). Although the highest accumulation of cadmium is found in bone, the liver and kidney also have a propensity for accumulating the metal, and the kidney is often seriously damaged in chronic occupational exposures (Lauwerys and De Wals, 1981). Clinically, patients suffer tubular dysfunction resulting in aminoaciduria, proteinuria, and glucosuria. Although the half-life of cadmium in kidneys of humans is uncertain, it may be as long as 30 years. Under such circumstances it has been conjectured that critical concentrations [kidney = 200 micrograms (μg) per gram (g) by age 50] could be used to establish maximum levels of daily exposure (Kjellström et al., 1977). What makes cadmium of dietary concern is that ordinary background dietary exposures were estimated to yield kidney concentrations of about one-quarter the hypothesized critical level. The segment of the population at greatest risk would appear to be older adults (ATSDR, 1989b). Studies in maternal-fetal tissues have provided evidence for accumulation and transplacental transfer of metals. In one study, placental cadmium levels were one to two times those in maternal or cord blood. It was observed also that erythrocyte cadmium levels were roughly three to five times plasma cadmium levels, and that maternal erythrocyte cadmium levels were somewhat higher (27%) than those of the fetus.

Lead

Of all the heavy metals, lead probably has the longest history of environmental contamination and toxicity to humans (Green et al., 1978). For this reason, lead poisoning, or plumbism, has been intensely studied, and a large body of information is available for examination. Sources of lead found in the environment are multiple, and the metal is truly ubiquitous, being commonly found in food, water, and air. Evidence exists that lead in the environment has increased during the past 200 years, and it is not surprising that it can be found as a contaminant of aquatic animals (Shukla and Leland,

1973). Environmental lead is a product of storage battery, ammunition, solder, pigment, pipe, brass, and red lead manufacture. Tetraethyllead is a component of gasoline antiknock additives, although in recent years this use has been drastically reduced. There are at least five pools of lead in the body, two of which reside in the skeleton (90%) in cortical and trabecular bone. Lead in cortical bone is similar in half-life to cadmium (approximately 20 years). Other body compartments for lead include the kidney, lung, and central nervous system (Goyer, 1986). It is not surprising therefore that major lesions and clinical signs in humans suffering frank plumbism are referable to the blood (anemia), brain (convulsions, paralysis), and kidney (proteinuria).

The condition in humans is best known because of its chronic toxicity to young children who ingest lead-base paint chips or lead in soil, house dust from paint, industrial dust, and automotive emissions. Oral ingestion of inorganic lead is without doubt the primary port of entry into humans. Of the lead ingested, only 5-15% is absorbed in adults but considerably more in children (Goyer, 1986).

Recent studies suggest that very low levels ingested by pregnant women may result in learning and behavioural disabilities in neonates and preschool children (Waternaux et al., 1989). Excretion is primarily via the bile and the gastrointestinal tract. Organic lead compounds such as tetraethyllead may be absorbed in large quantities through the skin, but as toxicants these forms of lead are primarily a problem in the petroleum industry. All forms of lead toxicity are less frequent in adults; any occurrence is usually acute and occupationally related (Green et al., 1978).

Lead's toxicological mode of action depends on its molecular configuration, inorganic lead being less toxic than and producing clinical signs different from tetraethyllead. Inorganic lead is an inhibitor of aminolevulinic acid dehydratase (ALAD) and heme synthetase, which leads to anemia (Hammond, 1978).

The metal causes necrosis of neurons, myelin sheath degeneration, and especially, brain vascular damage with increased cerebrospinal fluid (CSF) pressure. These lead to encephalopathy and eventual mental retardation in children. Lead crosses the placental barrier, and there is a good correlation between maternal and fetal blood lead values (Van Gelder, 1978). Therefore, at primary risk from contaminated seafoods are the fetus and neonates.

Mercury

Mercury exists in elemental form, as monovalent (mercurous) or divalent (mercuric) salts, and methylated. The methylated form is the most toxic to humans (Harada, 1978). Methylmercury is formed in the environment from the divalent salts by anaerobic bacteria. It is quite easily absorbed after ingestion and has a variable half-life of 60-120 days in man but is reported to have a half-life of up to 2 years in fish, where it is the predominant form (Al-Shahristani and Shihab, 1974; Stopford and Goldwater, 1975).

The metal is known to produce c-mitosis and chromosomal alterations resulting in cellular damage, with the kidney and brain as target organs. Neuronal damage and axonal demyelination result in the clinical signs and symptoms of paresthesia, incoordination, tremor, and epileptic seizures. The metal also binds strongly to sulfhydral groups (mercaptans), thereby inactivating certain enzymes (Hammond, 1978).

In its methylated form, mercury quite easily passes the placental barrier, placing the fetus at particular risk (Amin-Zaki et al., 1979). The relationship of clinical signs in humans to blood, hair, and urine mercury levels has been reviewed (Tollefson and Cordle, 1986). Children of symptom-free pregnant and nursing mothers with relatively low blood and hair levels may suffer from mental retardation.

Selenium

Selenium is an enigmatic metal because it functions both as an essential nutrient and, at slightly higher levels, as a poison. It is present in various enzymes, has been reported to possess anticarcinogenic effects in animals, is an antioxidant, and yet is a well-documented toxicant of domestic animals as well as a mutagen (Griffin, 1979; Schamberger, 1985; Schnell and Angle, 1983). As an animal toxicant it is a regional problem of the Southwest and Far West. Seleniferous (alkaline, oxidizing) soils give rise to high levels in selenium accumulator plants that are grazed by cattle, sheep, horses, and swine. Poisoned animals develop conditions known as "alkali disease" [subacute, <50 parts per million (ppm)] and "blind staggers" (acute, >100 ppm). Signs include anorexia, tooth and hair loss, watery diarrhea, lassitude, progressive paralysis, and eventual death (Harr and Muth, 1972).

Selenium levels in water from seleniferous areas are often quite high so it is not surprising that selenium has been found as a contaminant of fresh and marine aquatic animals. Its source however is not solely natural. Anthropogenic contamination occurs and is the product of fossil fuel combustion (fly ash) and of paint, alloy, photoelectric battery, and rectifier manufacture (Fishbein, 1983; Sorensen et al., 1984).

Selenium exists in a number of chemical forms, elemental selenium (Se^{o}), selenide (Se^{2+}), selenite (Se^{4+}), and selenate (Se^{6+}). These forms may bond with other metals or organic substances such as amino acids (Ewan, 1978).

The selenates are most soluble and easily enter biological systems. In one study, approximately 15-30% of the selenium found in fish muscle was the selenate form (Cappon and Smith, 1981). The selenites and elemental selenium are relatively insoluble.

This is not to say that selenite when ingested will not act as a toxicant, merely that its innate insolubility may affect its absorption and distribution within the body (Goyer, 1986). The biochemistry of selenium is poorly understood but has been reviewed recently (Reddy and Massaro, 1983). The mode of action of selenium as a toxicant at the cellular and biochemical levels is uncertain. The metal appears to damage endothelium selectively, resulting in edema and hemorrhage in both humans and animals. It is also responsible for toxic hepatitis with eventual fibrosis (not constituting cirrhosis) in chronic exposures. Selenosis in animals is reported to produce infertility and congenital malformations (Harr, 1978). Selenosis in man appears to be a relatively rare occurrence, most often due to acute occupational exposure or chronic exposure to contaminated water or food sources. There appears to be very little information regarding the effect on man of chronically high levels of selenium in the diet and its potential risk (Wilber, 1983). Recently, however, levels have been reached in fish that have prompted health alerts in California (Fan et al., 1988).

Organic Compounds

In this section, some of the potential contaminants of seafood that have come to the committee's attention, and about which there are at least some minimal data, are surveyed. These include the chlorinated hydrocarbon pesticides that came into widespread use in the United States and elsewhere immediately after World War II (Hansen et al., 1985). Among the chlorinated hydrocarbon pesticides detected in seafood were benzene hexachloride (BHC) or hexachlorobenzene (HCB), chlordane, dieldrin, dichlorodiphenyltrichloroethane (DDT), endrin, heptachlor, lindane, nonachlor, octachlor, and pentachlorophenol. In addition, industrial chemicals and by-products such as PCBs and dioxins are routinely detected in seafood. Less frequently detected pesticides included chlorpyrifos, dacthal (DCPA), diazinon, ethylene dibromide (EDB), malathion, mirex, omethoate, pentachloroaniline, tecnazene, and trifluralin (FDA, 1988; Gunderson, 1988).

In quite specific circumstances, such as in farm ponds in heavily agricultural areas, other chemicals — even those that are not known to bioconcentrate, such as atrazine — can be found in fish (Kansas DHE, 1988). Some pesticides detected are specific to various regions. The carboxylic acid herbicide 2,4-(dichlorophenoxy)acetic acid (2,4-D) has been found in oysters from the northern Chesapeake Bay and Alaskan bivalves (NOAA, 1988). Fish from the Arroyo Colorado and adjacent lower Laguna Madre in Texas contained measurable concentrations of pesticides such as ethion, carbophenothion, ethyl parathion, and methyl parathion (NOAA, 1988). The organic compounds classified here have been reviewed by Murphy (1980).

Specific Organics

Polychlorinated Biphenyls (PCBs)

Polychlorinated biphenyls include more than 200 different compounds ("congeners") that were used in various formulations as liquid insulators in electrical equipment, as encapsulating agents, in carbonless carbon paper, and in hydraulic fluids. The use of PCBs in "open" applications such as carbonless carbon paper was phased out in the early 1970s, and any new use for the remaining applications was stopped in the late 1970s with the passage of the Toxic Substances Control Act. The U.S. usage of approximately 500,000 tons of PCBs in 1930-1970 accounted for about half of the total world production. However, the unusually slow rate of environmental degradation of the more highly chlorinated PCBs in the environment and in higher organisms, and slow continued discharge of PCBs from old equipment and dump sites, have led to a relatively slow rate of decline of PCB concentrations in fish from large freshwater bodies (e.g., the Great Lakes).

The PCBs are a paradigmatic case for the phenomenon of bioconcentration. The more highly chlorinated congeners in particular tend to be both highly lipophilic and very slowly degraded by most organisms. Thus, PCBs that are passed "up" the food chain tend to become much more concentrated as predators are consumed by successively larger predators. In contrast, terrestrial animals that are used for human food are generally vegetarians (first-level consumers of the primary producing organisms).

The principal potential health concerns from PCB exposure include carcinogenesis (on the basis of extensive animal evidence and some suggestive findings in human epidemiological studies), changes in human birth weights, and some loss of neurological performance in the offspring of mothers with relatively high dietary exposures or body burdens. All carcinogens–in particular, the PCBs and dioxins–are not thought to act primarily by causing DNA mutations (Safe, 1989). Suffice it to say here that lack of knowledge of the precise mechanisms by which PCBs cause cancers makes quantitative assessment of their cancer risk more uncertain than is usual for other chemicals. The PCB mixtures that are delivered to humans via seafood are likely to be systematically different from the original mixtures that were used in animal testing because of more rapid degradation of some (particularly less chlorinated) congeners in the environment and in aquatic organisms. The selection for relatively persistent congeners in aquatic organisms might tend to increase human risk relative to that expected from a naive extrapolation; other factors might have the opposite effect. In any event, the numbers of cases that could be expected seem large enough to warrant exploration of further options for risk reduction.

Dioxins

2,3,7,8-Tetrachlorodibenzo-*p*-dioxin (hereafter known as TCDD) is a contaminant of products made from trichlorophenol, including some chlorophenoxy herbicides. In humans, its effect has been linked to a severe dermatitis; fetal toxicity and numerous other effects have been observed in experimental animals at very low doses.

In standard animal test systems, it is one of the most potent carcinogens known. Using its standard procedures for cancer potency estimation and certain consumption estimates, EPA estimated a lifetime cancer risk of approximately 1 in 100,000 from eating fish contaminated at the nominal detection level of 1 part per trillion (EPA, 1987). Using considerably different methodology for assessing the risk, FDA has advised that, for consumption patterns and species typical to the Great Lakes area, fish consumption should be limited if concentrations in the edible portion exceed 25 parts per trillion and should be banned if concentrations exceed 50 parts per trillion (Kociba et al., 1978). These profound differences in risk assessment indicate the tremendous uncertainty about the true potency of TCDD to cause human cancer.

Polycyclic Aromatic Hydrocarbons (PAHs)

Polycyclic aromatic hydrocarbons are common environmental contaminants found in petroleum, soot, or tar from incomplete combustion, lubricants, and domestic sewage. Many are well-established carcinogens and are highly toxic. Their pervasiveness in the environment assures widespread contamination of aquatic organisms. Because they are poorly metabolized by bivalves and are easily accessible to these animals, they may pose important potential hazards to humans.

Chlorinated Hydrocarbon Pesticides

Like PCBs, the broad group of relatively lipid-soluble, persistent chlorinated hydrocarbons was largely phased out of production in the United States during the 1970s because of concerns for carcinogenicity and ecological effects. Fortunately, few members of the group have proved to be as persistent as PCBs in the environment. Like PCBs, however, the precise mechanisms of action of many chlorinated hydrocarbons in causing cancer appear not to be by direct or indirect reactions with DNA; accordingly, quantitative assessments of risk for this group are more uncertain than usual.

DDT and metabolites: Both DDT and its metabolites [primarily dichlorodiphenyldichloroethane (DDE)] are persistent (slowly eliminated from organisms) lipophilic substances of uncertain health significance in humans, and are among the most widespread and frequently sampled of the chlorinated hydrocarbons.

They are also persistent in ecosystems and bioaccumulate at higher levels of the food chain, resulting in toxicity to birds and aquatic organisms. The use of DDT was essentially banned in the United States in December 1972. Subacute effects of these chemicals at high doses in humans include central nervous system signs and, in rodents, liver toxicity and estrogenic effects. In addition, DDE has been observed to cause liver tumours in rodents.

Dieldrin: Dieldrin (an epoxide of aldrin) is a cyclodiene insecticide that, like DDT, affects the central nervous system, but is more toxic and has caused human fatalities. It too is lipophilic and may be released from fat stores long after exposure, to cause toxicity. It has led to increased liver tumours when fed at relatively low levels to rodents.

Chlordane: Chlordane is similar in molecular structure and mode of action to dieldrin, but is less toxic.

Heptachlor and heptachlor epoxide compounds: Heptachlor and heptachlor epoxide are also chlorinated cyclodienes, and the epoxide is known to be stored in human fat. They have toxicity similar to dieldrin.

Endosulfan: Endosulfan is a cyclodiene pesticide and a problem contaminant in estuaries near agricultural drainage areas due to its widespread use (NOAA, 1989).

Endrin: Endrin is similar in its toxic effects to other cyclodiene pesticides and is more acutely toxic than DDT.

Pentachlorophenol: Also known as PCP and penta, pentachlorophenol is a wood preservative, slimicide, and metabolite of the fungicide hexachlorobenzene. Like other polychlorinated phenols, it is contaminated with carcinogenic dioxins (NOAA, 1988).

Mirex: Mirex is a pesticide used to control the fire ant in the southeastern United States. Like other organochlorine pesticides, it is lipophilic and has been reported to be a carcinogen on the basis of rodent studies. It may be a precursor of chlordecone (kepone), is persistent in the environment, and bioconcentrates in the food chain.

Kepone: Kepone has produced appreciable toxicity in exposed workers. It can cause neurological lesions, liver damage, and reproductive failure and is similar in its bioconcentration properties to mirex.

Toxaphene: Toxaphene is a very common domestically used insecticide of complex and often uncertain molecular structure. It is made by chlorinating a mixture of terpenes. It therefore may vary in toxicity from batch to batch depending on the proportion of its isomers. Fortunately, it is of relatively low persistence in the body. Carcinogenic activity is suspected.

Carboxylic herbicides: The herbicides DCPA, 2,4-D, and 2,4,5-trichlorophenoxyacetic acid (2,4,5-T) are chlorophenoxy compounds commonly used in agriculture, and by government agencies and utility companies to control woody plants in rights of way and along roadsides.

Although 2,4,5-T has been found to have teratogenic activity, there has long been discussion about whether this is attributable to small amounts of dioxin contaminants. Aside from teratogenesis, these agents can affect animals by impairing neurotransmission, resulting in muscle weakness, ventricular fibrillation, and neuritis.

Atrazine: Atrazine is a herbicide commercially known as AAtrex. It is of low oral acute toxicity.

However, its chronic effects are unknown. Structurally similar compounds have been shown to produce thyroid adenocarcinomas in rodents.

Contamination Problems in Aquaculture

Fish culture uses a variety of chemicals that represent potential threats to the health of the cultured animal, indigenous biota, and even the human consumer (Meyer and Schnick, 1989). A number of chemicals of potential toxicity to humans that are not registered for use in the United States are employed in other nations (Fox, 1990). These include furazolidone, nitrofurazone, carofur, chloramphenicol, and silvex–all of which are known or suspected carcinogens.

Chemicals employed in aquaculture include (1) drugs used to treat disease (chemotherapy), (2) those introduced through construction materials, (3) chemicals to treat parasites (formaldehyde), (4) hormones used to alter reproductive viability, sex, and growth rates, and (5) water quality treatments (copper compounds). Of these groups, those of greatest potential concern are the chemotherapeutic drugs. Chemicals used in construction and hormones are not considered in this section because they are relatively nontoxic or have been considered under other headings (organics, pesticides).

Disease is a limiting factor in the culture of aquatic animals. In recent years, culturists have dealt with this problem by developing rapid, immunological diagnostic tests, followed by treatment with drugs commonly used in veterinary medicine. Such chemotherapeutic drugs include, but are not limited to, the potentiated sulfonamides, antibiotics, and nitrofurans. Their widespread use, both nationally and especially internationally, may pose a threat to human health if residues persist in the edible portions of fish and shellfish. Both residue tolerance levels and withdrawal times (the period prior to slaughter during which no drug may be administered) have been determined and set by FDA for major drugs used by veterinarians in the treatment of terrestrial food producing animals. Hence, the problem of persistent residues has long been recognized by the medical profession, but only recently has it been investigated in cultured aquatic animals. The Centre for Veterinary Medicine of FDA is aware of the seriousness of this problem and is actively attempting to deal with it, as evidenced by its recent requests for proposals in the April 9, 1990 issue of *Business Commerce Weekly*.

The proposals will study the pharmacokinetics and methods of monitoring some six drugs, including malachite green and chloramphenicol, presently used in fish and crustacean culture. In addition to this organization, active research is being conducted by the Fisheries Research Branch of the FDA and the Fisheries Research Centre, La Crosse, Wisconsin of the U.S. Fish and Wildlife Service (Schnick, 1988). The latter organization has published a booklet entitled "A Guide to Approved Chemicals in Fish Production and Fishery Resource Management" (Schnick et al., 1989). Although there is much interest in the potential toxicity of chemotherapeutic drug residues in seafood, to date only a potentiated sulfonamide (Romet-30), oxytetracycline (Terramycin), and Formalin are approved for use in food producing aquatic animals. The nature, use, and residues of the drugs applicable to aquaculture that are used to control disease in veterinary medicine have been reviewed. Chemotherapeutic drugs used in aquaculture are sulfonamides, antibiotics, and drugs used in the chemotherapy of protozoal, mycotic, and helminthic infections. In addition to these examples of deliberately used chemicals, fish raised in aquaculture are also susceptible to contamination via pesticides present in feed, agricultural runoff water, and sediments. The magnitude of human exposure to these sources has not yet been assessed and should be examined periodically in light of the growth and change in this sector of the seafood industry.

Antibiotics

The term antibiotic includes a large number of drugs inimical to the growth of microorganisms. Of greatest interest in aquaculture are oxytetracycline, ampicillin, chloramphenicol, and erythromycin.

Oxytetracycline (Terramycin) is a product of the mold *Streptomyces rimosus*. It is a broad-spectrum antibiotic used to treat a number of bacterial diseases in fish and crustaceans. It is the only antibiotic approved by the FDA for use in aquaculture. Contamination of the food supply with the drug is a public health concern because chronic exposures at low levels may lead to a higher incidence of antibiotic-resistant bacterial strains, poor growth of teeth, and the possibility of photosensitivity. The FDA tolerance level for oxytetracycline in meat is 0.25 ppm, whereas in fish it is 0.1 ppm with a withdrawal time of 21 days. Elimination time from muscle appears longer for salmonids than for catfish and is dependent on the water temperature (Plakas et al., 1988).

Ampicillin is a newer, semisynthetic penicillin that has activity against both gram-positive and gram-negative organisms. Although not approved for use in the United States and seldom used domestically, it is commonly used in Japan to control pasteurellosis in yellowtail culture (Hawke et al., 1987).

Ampicillin, like its congeners, may result in severe hypersensitivity reactions in some people (Huber, 1977a). Because an initial exposure to some form of penicillin is necessary to produce an eventual drug allergy, ingestion of ampicillin as a residue in seafood is of public health interest. The FDA has set a tolerance level of 0.01 ppm and a preslaughter withdrawal time in cattle of 6 days. No information is available regarding residues or withdrawal time in fish or crustaceans.

Chloramphenicol is a product of the mold *Streptomyces venezuelae* and is a potent antibiotic that is effective against most bacteria, as well as rickettsia and the psittacosis-lymphogranuloma group of organisms. Development of resistant bacteria may follow its widespread and uncontrolled use.

Chloramphenicol administered for long periods may cause blood dyscrasias such as aplastic anemia and has recently been incriminated as a carcinogen.

It is not approved for use in the United States in food producing animals. The drug is used, however, with impunity to control bacterial diseases of shrimp, especially in Ecuador and in European fish culture (Manci, 1990; Meyer and Schnick, 1989). Because imported seafood is not checked for antibiotics, no information is available regarding its residue levels.

Erythromycin is a product of the mold *Streptomyces erythreus* and is primarily effective against gram-positive organisms. The drug is not approved for use in domestic fish culture but is used in Europe. The tolerance level in U.S. swine is 0.01 ppm, and the withdrawal time is 7 days. The drug is not considered an important cause of hypersensitivity in man and is relatively nontoxic.

Sulfonamides

The sulfonamides are readily synthesized derivatives of sulfanilic acid. They are bacteriostatic drugs, and their efficacy may be increased by the addition of ormetoprim (Plakas et al., 1990). Commonly, sulfadimethoxine is potentiated with ormetoprim (Romet-30) and fed to cultured fish at the rate of 50 mg/kg body weight for 5 days. Similarly, sulfamethoxazole is potentiated with trimethoprim (Tribrissen) and fed at the rate of 1-4 mg/kg body weight for 10-14 days. The latter drug is not approved for use in aquaculture. The FDA tolerance level for both these drugs in cattle and chickens is 0.1 ppm (Booth, 1977). The sulfonamides have been implicated in renal damage, urinary obstructions, and hematopoietic disorders. For that reason, FDA has set withdrawal times and residue tolerance levels to protect public health. The withdrawal period for catfish is 3 days and for trout 6 weeks (Schnick et al., 1989).

Nitrofurans

Nitrofurans are synthetic compounds that are active against most gram-negative bacteria, some fungi, and some protozoal organisms. Because of their toxicity, their clinical use is limited and FDA has attempted unsuccessfully for over 15 years to ban their use in food producing animals and to have them withdrawn from the market (FDA, 1982). Side effects include bleeding, gastrointestinal upsets, and allergic reactions. Tolerance levels for the drug have been set at zero. In aquaculture, their use has been suggested for the treatment of *Ichthyophthirius multifillis*.

7

Remote Sensing to Fisheries

Remote sensing techniques can be utilized directly, indirectly or as general aids in the detection and assessment of fishery resources.

The sea covers two thirds of the earth's surface. To a large extent, man is dependent on it for food species which include fish, shellfish, marine mammals, turtles, aquatic plants and algae. To exploit these resources more effectively, fishermen must catch the most fish possible (within biological constraints) while, at the same time, minimizing costs and optimizing the scheduling of their operations. Reliable environmental information is required from the scientific community for these purposes. Remote observations of the sea surface can provide a significant part of the information needed to assess and improve the potential yield of the fishing grounds. In the past, remote sensing was used predominantly to assist in the efficient harvesting of natural resources. Today it is being used for resource management, conservation and exploitation.

Variations in environmental conditions affect the recruitment, distribution, abundance and availability of fishery resources. It is not possible to measure remotely the entire range of information needed to assess changes in the marine environment. Knowledge of particular conditions and processes affecting fish populations, however, may often be deduced using measurements made by remote sensors, e.g., concentration of dissolved and suspended matter, variations in primary production levels, distribution of surface isotherms, location of frontal boundaries, regions of upwelling, currents and water circulation patterns. The parameters providing information on these environmental factors may allow a forecast of fish distribution or more generally the definition of marine fish habitats. These are often easier to sense remotely than the presence of fish.

DIRECT METHODS OF FISH DETECTION

The most direct and simple method of remote sensing in fisheries is visual fish spotting. Fishing fleets which exploit major fisheries such as tuna and menhaden are dependant on visual fish spotting from aircraft to direct their fleets.

Aerial photography per se is of little importance to the majority of commercial fisheries. The location of mobile fish schools, for example, cannot be provided fast enough to the fishermen. Aerial photography, however, can be of assistance to a fisheries scientist as it provides information about the distribution and relative abundance of pelagic fish, particularly the schooling species. The pattern of distribution and the location may identify the species observed, and the surface area of a school, measured from an aerial photograph, has been shown to be correlated with the biomass of some species.

Echo-sounders and sonars have been in use as remote sensors for at least 50 years and are now widely used by the fishing fleets of the world. Sonars are useful for the detection of fish and biomass estimation.

In recent years, high powered laser systems operating in the blue-green portion of the visible spectrum (lidar) have shown promise for the evaluation of fishery resources. A lidar carried on aircraft flying at an approximate altitude of 1700 m can detect fish at depths to 16 m.

INDIRECT METHODS OF FISHERY ASSESSMENT

Estimation of a fishery resource can be assisted by the measurement of parameters which affect its distribution and abundance. Much of the research dealing with environmental effects related to fisheries are concerned with the correlation of a single parameter with the spatial and temporal distribution of fish. It is most likely, however, that fish respond to the sum total of environmental factors. Thus, it becomes necessary to correlate a large number of parameters, obtained by remote sensing techniques, with fish distribution.

The environmental parameters most commonly measured from airborne and spaceborne sensors are as follows: surface optical or bio-optical properties (diffuse attenuation coefficient, total suspended matter, yellow substance, chlorophyll pigments and macrophytes, commonly grouped under the general term of ocean colour); surface temperature; vertical and horizontal circulation features; salinity; oil pollution; and sea state.

Surface optical properties

The optical properties in the marine surface layer are determined by the presence of dissolved and suspended matter. Under normal conditions, visible light penetrates marine waters to a depth of tens of metres. As the concentration of the water constituents increases, i.e. the water becomes more turbid, the penetration of sunlight is reduced as a result of absorption and scattering processes. Depending on the specific characteristics of the materials present in the water, i.e. on their spectral signature, the absorption and scattering processes will vary with the wavelength of the incident radiation. Multispectral observations, therefore, can be employed to estimate the nature and concentration of the water constituents. Passive sensors working in the visible wavelengths (mainly CZCS but also MSS, TM, HRV) are commonly used to image water colour. Active sensors providing their own source of illumination, e.g., lidar, can also be used but only from aircraft and for sampling, rather than for imaging purposes. The main parameters which can be derived remotely from water emergent radiation, through the use of empirically constructed algorithms, are listed below.

Diffuse attenuation coefficient: The diffuse attenuation coefficient at a specific wavelength is an apparent optical property. Its magnitude depends on the light distribution as a result of spreading, scattering and absorption that exists at the in situ point of measurement. This parameter, when correlated with Secchi disk depth and Munsell colour hues, provides the means of physically categorizing water according to colour. Its value can be interpreted as a measure of water turbidity and it constitutes a valuable tool in fisheries studies. It has been shown, for example, that turbidity and menhaden sighting in the Mississippi Sound are highly correlated.

Total suspended matter (seston): In addition to optical parameters, the total concentration of the absorbing and scattering agents can be used to classify surface waters by means of their colour. The utilization of this parameter may be most appropriate when classifying waters where inorganic and/or organic sediments make an important contribution to the optical properties of the surface layer. It may also be appropriate if sediment concentration has to be used as a natural tracer for the identification of water movement and frontal boundaries.

Yellow substance: The term yellow substance may be defined as the material derived from the degradation of land and marine organic matter. It is an important parameter to monitor in the context of polluted coastal waters, since it may be used to identify marine areas where the exploitation of filter feeders, e.g., shellfish, could be hazardous. In certain regions of the world, for example the North Sea, this parameter exhibits some correlation with the salinity of surface waters.

Chlorophyll pigments: The concentration of chlorophyll pigments (the photosynthetic pigments of phytoplankton) is often considered as an index of biological productivity and, in an oceanic environment, it can be related to fish production. Chlorophyll concentrations above 0.2 mg/cu.m indicate the presence of sufficient planktonic life to sustain a viable commercial fishery (Gower, 1972). Chlorophyll pigments have a specific and distinctive spectral signature since they absorb blue (and red) light and reflect strongly the green, thus affecting ocean colour. Multispectral observations from airborne or spaceborne sensors, therefore, allow the deduction of phytoplankton concentration.

Macrophytes: In coastal areas it is common to find macrophytic vegetation (seaweed). Some species are of economic importance but all species play an important role in supporting marine life. Different kinds of seaweed have different light reflection properties, for example, reflect more green or red radiation. This distinction which allows the differentiation of some seaweed species can be detected from airborne or spaceborne passive visible sensors. Due to the low intensity of the light as it leaves the water, however, it is usually more effective to employ airborne sensors such as aerial cameras or radiometers.

Surface temperature

Since 1973, the US National Oceanic and Atmospheric Administration (NOAA) has been engaged in sea surface temperature (SST) determination from satellite derived data. The process of extracting SST information from IR radiometer data is well established. Global sea surface temperature charts are produced on an operational basis. They are in the form of computer printouts or contour maps with spatially smooth and radiometrically corrected measurements. It has been possible with data derived from TIROS, NOAA and the METEOSAT satellites to produce SST charts with an accuracy of 0.5°–2°C and in near real time.

The heliosynchronous satellites of the NOAA series provide high resolution (1 km) pictures twice daily while the geostationary satellites (GOES, METEOSAT) provide pictures every 1/2 hour but with a resolution of only 5 km. The geostationary satellites are principally used for the near-equatorial area where the sensor's resolution is at its best. For latitudes higher than 40° the image distortion is generally too extreme for operational use.

The occurrence of cloud or haze contaminates data to a certain extent but a knowledge of day-to-day variations or trends enables corrections to be made by interpolation. The sea truth information provided by ships is of further assistance in deducing the precise temperature fields.

To date, SST maps are mainly used by the salmon and tuna fishing fleets. It is well known that some tuna species feed on the warm seaward side of thermal fronts while salmon feed on the cold landward side. The occurrence of some other species can also be correlated with SST. In addition, physical features such as gyres, eddies, inversions and upwelling which are of importance to fisheries can be detected using SST maps.

Circulation features

Several remote sensing techniques can provide information regarding surface circulation features of importance in defining marine fish habitats. These include the location and evolution of frontal boundaries, upwelling areas, currents and circulation patterns in general. Optical and thermal characteristics of surface waters can be used as natural tracers of dynamic patterns. Hence, the previous discussion of sea surface colour and temperature should be considered again in light of this application. Microwave techniques, particularly the use of active sensors (radar altimeter), also have applications regarding large-scale circulation features. For example, remote measurements of water surface vertical displacements can provide information on the dynamic characteristics of a basin.

Salinity

The measurement of salinity from remotely sensed data is not operational at the present time. Research, however, indicates the possibility of determining salinity with the use of microwave sensors to an accuracy of one part per thousand. The microwave properties of the sea surface are a function of its physical and chemical state. The emissivity of sea water is related to salinity. Changes in salinity cause significant changes in the emissive brightness temperature of water for frequencies less than 5 GHz. Hence the salinity of sea water can be determined remotely by measuring accurately the emissive brightness temperature. The precision afforded by this technique may be adequate for mapping the spread of fresh water at a river mouth or for studying estuaries and near shore waters.

Oil pollution

The numerous methods used for oil detection at the sea surface include visual detection by eye, aerial camera, MSS and CZCS; microwave detection by SMMR and SAR; fluorescence detection by lidar; and thermal detection by IR scanner.

The visual method images the change in colour and brightness due to the presence of oil. Other visible-light phenomena used to detect oil slicks include EMR interference effects (colour banding) and the suppression of solar speckle by slicks. The microwave method, when passive techniques are used, is based on the difference of emissivity between the sea surface and the oil slick. Active radar sensors depend on small capillary wave backscatter to be dampened by the oil slick as a means of oil detection.

Fluorescent properties of hydrocarbons may be detected and discriminated by appropriate lidars. These laser fluorosensors can also identify the basic types of oil (heavy, light, etc.) and provide a measurement of oil slick thickness. Thermal sensors identify oil by means of the difference in solar absorption and thermal emissivity between oil and water and they also provide a basic measurement of oil thickness.

General Aids to Fishing Operations

Satellites can assist the fishing industry in many ways other than the locating of fish per se. Most of these aids are also of assistance to mariners other than fishermen. The types of assistance that satellites can offer include the following:

i) search and rescue operations: The satellite NOAA-8 carries a special sensor, SARSAT (Search and Rescue Satellite Tracking), which detects the distress signals emitted by vessels in difficulty. The recorded signal is used to locate the position of the vessel. Sensors carried on board the Russian satellite series COSPAS-1, 2 and 3, launched respectively in 1982, 1983 and 1984, fulfill a similar function to SARSAT;

ii) weather reports: Environmental satellites such as NOAA, GOES or METEOSAT can provide weather information over a wide area at a given time. This may assist fishermen to plan their fishing operations. In higher latitudes, ice and icebergs are major hazards; environmental satellites can assist in identifying ice and spotting icebergs;

iii) bathymetry: Remote sensing using passive or active visible sensors may be used for bathymetric measurements. With the exception of acoustic methods (sonar), airborne sensors provide the most accurate bathymetric measurements. In addition, active sensors such as bathymetric lidar are more reliable than the passive devices.

FISHERIES: AN OPEN ACCESS RESOURCE

Due in large part to the way it has traditionally been managed, marine fish has usually been seen and used as an open access resource, an in many ways endemic characteristic which has been the base of its dynamics of use, management, and exploitation.

The term refers to a situation where no single user has to pay for access to a resource, nor does that user have exclusive rights to the resource or the right to prevent others from using it. Physical characteristics of the oceans, and the fish they contain, such as scale, interconnectedness, and mobility of fish stocks, contribute to this open access nature. As such, they have usually defied national borders and been logistically difficult or impossible to monitor and govern. Theoretically, this can lead to a "tragedy of the commons," where "the open-access nature of the fisheries allows more and more people to enter the fishing sector, which leads to the economic (and possibly biological) over-exploitation of the resources."

Even after the resource begins to decline, producers will continue to increase their efforts until operating costs can no longer be covered. By this time, the stock of the resource may be seriously depleted. The common-access nature, which has long been a characteristic of fisheries, is the foundation on which many of the causes of current problems in the sector are based.

Root Global Causes of Fisheries Resource Depletion

The simultaneous, and highly linked, realities that too many fish are being caught and the global demand for fish is unsustainably high are largely seen to be the result of four root causes: population growth, growth in incomes, urbanization, and advancements in fishing technology.

Population growth has contributed to an increase in demand for fish by increasing the number of people who demand fish. Income growth has played a role because, as incomes increase, wealthier people's tastes change from preferring low-value to high-value fish; more and more low-value fish are then diverted "from poor people's plates into the stomachs of expensive carnivorous fish, which will in turn end up on the plates of wealthier people." A 2003 report by the World Fish Center found that population growth and income growth were "widespread structural phenomenon," especially in the developing world, which drove increased demand and consumption in the last three decades.

Urbanization has also impacted the structure of fish demand because, as the "rapidly expanding" group of poor people living in larger cities throughout the world increases, low-value fish is their only affordable source of animal protein. On the production side, developments in technology that have allowed fishermen to catch more per unit input and seek out fish in deeper waters have kept production costs down and allowed production levels to increase.

Roman Grynberg and Martin Tsamenyi, for instance, cited constantly improving technology as a fundamental cause of fish stock depletion. The interaction of these structural forces of increasing incomes, populations, and urbanization, as well as advancements in technology, have lead to the over-use of the resource, which is, as is currently being witnessed, limited in its capacity to respond to and accommodate them.

MANAGEMENT AND SUBSIDIES

The very reality that the root causes of the over-use of fish stocks are structural makes it nearly impossible to completely alleviate the pressures of rising demand, increasing abilities to pay for high-value species, and more effective ways to capture fish on an open-access resource. But people, organizations, and governments have responded to these realities in many ways, their responses sometimes improving the situation, but often exacerbating it. Two of the most prominent and important responses have been subsidies and management schemes. The first, subsidies, occur when trade measures directly influence the income or reduce the costs of production for industry. Management schemes, the second, encompass the ways in which fisheries are governed and monitored. In practice, the way this is done often highlights a lack of effective management. Subsidies and management, which can be seen as responses by individuals, groups, or governments to the root causes of fisheries depletion, are also the areas where the most opportunities for change in the way fisheries resources are used exist.

Effective management regimes are considered by many to be key in the sustainable use of fisheries resources. Theoretically, effective management encourages a sustainable use of resources through providing conservation regulations and incentives for responsible use of the resource. They are especially beneficial under the pressures of international integration; responses by producers to the economic incentives from trade liberalization, for example, can be better regulated when effective management schemes exist so that over-exploitation and destruction are less likely to occur. Unfortunately, the prevalence of such management has been scant in real-world fisheries. As the United Nations Environment Programme (UNEP) reports, "very few fisheries management systems have demonstrated the ability to keep catches below levels that put pressure on stocks."

As noted in a study by the World Conservation Union (IUCN), "national and international progress on trade expansion and liberalization outpaces progress on fisheries management and articulation of sustainable development strategies." Fisheries management, crucial to sustainable use of the world's fisheries, has been difficult to implement effectively and has often left fisheries open to influences from other sources such as state trade measures.

Subsidies, one of these trade measures, have played a proven role in aggravating the causes of fisheries depletion. Theoretically, subsidies can lead to overuse of a resource by altering the behaviour of producers. Revenue-enhancing or cost-reducing subsidies, such as those that will be examined in this paper, increase marginal profits at each level of effort and therefore theoretically lead to an increased overall effort. This may have the short-term effect of creating additional economic rent for producers, but either new entrants or increased effort by existing producers, also encouraged by the subsidy, will shift the level of effort so that rent is eventually dissipated. Assuming that the management program does not effectively impose a sustainable level of resource use, cost-reducing and revenue-enhancing subsidies will push the level of overcapacity and overall effort even further than would an open-access resource in the absence of subsidies. The main theoretical effects of cost-reducing, rent-increasing subsidies are increased capacity, a delay in exit by producers from the sector, and increased effort by producers. These effects can cause the overuse of a resource, which can have many other significant impacts such as threats to food, livelihood, and resource security.

FISHING SUBSIDIES

Types Used by National Governments

In the real world of governments, fishers, fisheries resources, and the incentives that link them all together, a subsidy can take many forms. Fishing subsidies currently account for large portions of many industrialized country budgets; a 1998 World Bank study estimated annual budgeted expenditures on global fisheries subsidies to be between $14 and $20 billion, and that subsidies account for 20 to 25 percent of the sector's revenues. One type of subsidy, domestic assistance programs, form a clear example of government endorsement of the fishing sector. These programs are defined as budgeted assistance to fishing production, and often enhance operations and capacity. Japan, one of the biggest subsidizers of fishing activities in the world, has used domestic assistance programs to recruit young fisherman, give aid to fish cooperatives and boat owners, support marketing, and encourage price stabilization.

A 1998 World Bank report estimated that Japan budgets $270 million a year for these types of programs, and a total of $750 when aid for fishing vessel insurance is included.

The European Union (EU), another major subsidizer, supports such structural programs as fleet renewal and modernization, processing and marketing, aquaculture, maintenance of port facilities, and genetic product promotion in their fisheries sector. In addition, it pays for market supports to the sector in the form of a minimum import price program and measures to support price floors. Costs of these programs to the EU in 1996 were estimated to be about $530 million. While estimates of exactly what types of domestic subsidy programs the Chinese government uses are hard to determine due to scant data, they have clearly aggressively promoted the expansion of their fishing sector in the last 15 years (and have become the world's leading producer) and are estimated to budget between $500 and $750 million annually for domestic subsidies. While other countries subsidize their domestic fishing industries, they have comparatively small budgets for such programs as so do so on a smaller scale. The shear magnitude of the domestic subsidies in industrialized countries suggests the possible impacts on the behaviour of fishers their presence may cause.

A less obvious, but yet potentially just as influential, type of subsidy occurs when a country pays the access fees for their fleets to fish in foreign waters.

Unlike domestic subsidies, these payments are not made directly to fishers or even domestic programs, but to the foreign governments who have jurisdiction over the fish stocks. The EU is one of the major subsidizers of distant-water fishing through its numerous access agreements with developing countries, mostly in West Africa. For example, the EU signed a five-year access agreement with Mauritania in 1996 that lifted an EU embargo on fishery imports from Mauritania and specified EU payments of almost $350 million over five years to the Mauritanian government. In return, the Mauritanian government granted increased access for EU fishing vessels in their waters, authorization of higher EU total harvests, and the specific allowance of EU-directed fishing of highly valued squid and octopus.

The monetary benefits to the "cash-strapped" nation were described as a "windfall;" in consequence, the number of eligible EU boats rose from 165 to 240 and allowable EU harvests from 76,050 to 183,392 tons. The EU has similar agreements with other West African nations, as well as others in East Africa, the Indian Ocean, and South America. Another distant-water fleet subsidizer, Japan, currently budgets about $200 million a year on distant water arrangements with foreign countries and foreign fisheries assistance programs (which often help secure fishing rights). These payments went mostly to developing countries in the Pacific Ocean and elsewhere. The subsidization of distant water fleets from industrialized nations can have serious effects on the coastal communities of the developing countries with whom these agreements are typically made.

Possible Impacts of Perverse Subsidization

The impacts of government subsidies of fishing are at least as complex and numerous as the types of subsidies themselves, but in the context of sustainable development those related to market access due to trade measures and resource, livelihood, and food security are the most relevant. The trade implications of the subsidization of the fishing industry arise when the lower costs producers see as a result of the subsidies repress world prices for fish products. Those producers, who are often from developing countries that cannot afford subsidies, are then excluded from the market because they cannot afford to produce at the world price.

For example, when Japan subsidizes its fleets to fish in the waters of Chile, Japanese fishers bring the fish back to Japan and sell them on their domestic market. Chilean fishers try to enter the Japanese market and sell these same Chilean fish to Japanese consumers, but are kept out of the market because their costs are higher and the prices on the Japanese market are too low. In effect, Japanese fishers push Chilean producers out of the market for the sale of their own fish. This example illustrates the way in which subsidies in the fishing sector can exclude those who are outside the national subsidy regime from participating in the market.

Due to numerous examples where the presence of subsidies seemed to increase the capacity in the sector and, in turn, deplete available resources, fishing subsidies have also been blamed for a tendency to encourage overcapacity and overfishing. The case of stock depletion in Canada's Northwest Atlantic fisheries in the 1950s and 1960s illustrates how subsidies can increase capacity and lead to depletion. Between 1954 and 1968, Canadian subsidies increased the capacity of the Northwest Atlantic offshore fishing fleet by more than 18 times, creating twice as much capacity as could capture fish if resources were to be used sustainably.

In 1970, the Canadian government "acknowledged explicitly that its subsidies for vessel construction and modernization over the previous two decades had led to the rapid expansion of larger vessels, creating serious overcapitalisation." An Organization for Economic Cooperation and Development (OCED) study reported that subsidies also contributed to overcapacity in the fishing sector in many member states: New Zealand in the 1970s and early 1980s, Spain's Galician fisheries in the 1980s, Norway in the 1960s, the EU during the 1980s, and the United States (US) in the late 1970s and 1980s. These examples of the history of fishing fleet subsidization in industrialized countries and the resulting depletion of their stocks illustrates how overcapacity in a sector can encourage overuse of a fragile resource and lead to its exhaustion, posing threats to the surrounding ecosystem and human environment.

In addition to threatening the resources on which humans depend for their incomes, fishing subsidies may pose direct threats to crucial aspects of human and national development such as livelihood and food security. These threats are seen most strikingly in the interaction between developing nations and the industrialized countries with whom they sign access agreements. Frequent realities of EU-West African access agreements, for example, include a disregard of African fishing regulations and international agreements by distant water fleets, an inability by African states to monitor activities or catch levels of EU fleets, a lack of direct linkage between access payments and money spent on fisheries-related activities, a lack of enforcement of management efforts by African governments due to a fear of losing compensations, and the challenge of depletion of valuable fisheries stocks.

There are also often many adverse effects on African coastal communities such as threats to the livelihoods of artisanal fishers and food security of those who depend of local fish stocks for protein intake. While EU payments often constitute important sources of revenue for West African governments, the security of fishing resources, vital to so many aspects of these countries, are often threatened.

The case of one West African country that signed an access agreement with the EU, Senegal, illustrates how subsidies and trade agreements can lead to threats on food and livelihood security. The Senegalese government initially implemented subsidies as a way to encourage fish production to supply domestic consumers with a protein source; fish accounts for 75 percent of the population's protein needs. But the use of subsidies has since become incompatible with this goal since the end result has been a re-orientation towards encouraging the exportation of fisheries resources and agreements with the EU for access to Senegalese fisheries.

The combination of domestic export subsidies, which lower the costs of production for those fishers who chose to export, and trade advantages granted under the Lomé Agreement of 1982, which authorized customs exemptions to products originating in African, Caribbean, and Pacific (ACP) countries and increased the competitiveness of Senegalese piscatorial products in the European market, have encouraged small-scale fishers in the country to both intensify their efforts in response to international demand as well as orient themselves towards the export market.

These developments in subsidy policy, along with several fisheries agreements that have been signed with Japan and the EU, have threatened livelihood and food security in Senegal. The impact of growth in subsidy-driven exports has lead to a "breakdown in the supply of cheap protein to the population," since local producers began to switch their activities toward the capture of species of high market value in European economies rather than those affordable to local consumers.

The increasing scarcity of fisheries resources has also affected the traditional, artisanal fishers, whose techniques, which require low technology and employ many people, are beneficial for local resource and employment security. Since fish have become scarcer, foreign fishing fleets have become a larger presence, and local fishers lack the equipment and fuel to travel farther out to sea where the remaining catches are located, they have been forced to make deals with European or Asian boats to remain employed.

In return, locals allow the foreign boats access to areas originally reserved for their needs and use. This trend threatens local control of the resource as well as livelihood security. In a country where fish provides a vital source of sustenance for the local population, in terms of both food and employment, the way in which subsidies encourage a reorientation of resources to foreign markets is a matter of much concern for the sustainable development of the country.

8

Fisheries Development in Biotechnology

The gap between human demand and the availability of livestock products is huge in Africa, and imports have been increasing fast. The productivity of African breeds is very low: mature beef cattle of 4-5 years hardly weight 300 kg (as against 400 kg at one year for exotic animals); the best African cows produced 300 litre/lactation (as against 5 000 litres and more in Europe); 100 African ewes produced 50 lambs every two or three years (as against 150 lambs a year in developed countries).

Retrospectively, production policies that aimed at developing livestock through cross-breeding of exotic grade cattle with indigenous ones have generally failed. The new approach is not to make selection of local performant animals. Biotechnology offers good methods both for selection and cross-breeding schemes, essentially through Multiovulation Embryo Transfer (MOET) and the field diagnosis of pregnancy. Yet, there is wide diversity among the countries to use new technologies. For instance, even if simpler techniques such as Artificial Insemination (AI) are considered, it is seen that in East Southern Africa the technique is widely used, whereas in West and Central Africa, because of the predominance of nomadic transhumant livestock production, the technique is least developed.

MOET is well developed in few countries such as Zimbabwe and South Africa. Cattle farmers in these countries import frozen embryos from overseas. MOET is promising especially in the development of dairy cattle farms around large cities. Both AI and MOET are being used in Open Nucleus Breeding Systems (ONBS) to improve animal production. Small-scale ONBS projects are being carried out by FAO in small ruminants in Ghana and the Gambia. Moreover, by using semen from bulls, with genetic resistance to high temperature, diseases and insects common to African countries, embryos could be produced from superior cows of other regions and then implanted into surrogate mothers within the environment where improved livestock are needed.

In animal feeding, biodegradation of low-quality forage is also a promising technology. It can be associated with chemical degradation techniques (use of urea and ammonia) to improve animal feeds during dry seasons.

In some of the countries, farm diagnosis of pregnancy is also being used by field workers by using a kit based on the difference of levels of progesterone in cow's milk.

In the field of animal health, Sterile Insect Technology (SIT) is well known in Africa. Sterile males of glossina are being reared in Bobo-Dioulasso and Burkina Faso for field application. Successful field projects in tsetse control have been carried out in Nigeria (BICOT Project) and in Burkina Faso. Newworld screwworm was eradicated in the Libyan Arab Jamahiriya in 199091 by the use of SIT.

In disease diagnosis, ELISA is being used to evaluate rinderpest immunity in vaccinated animals. Monoclonal Antibody Techniques (MAB) is not commonly used. The two techniques are very promising because they are very simple and most effective. Trials are under way to use the ELISA test in trypanosomiasis diagnosis. In vaccine production recombinant rinderpest vaccine trials done in the United States seem ready to be utilized in a mass vaccination campaign against rinderpest in Africa. There is an increasing level of activity in hybridoma work in sub-Saharan countries.

The work of ILRAD in Kenya is addressing the problem of diseases and parasites that affect cattle. Some of the common cattle diseases are East Coast Fever, tickborne diseases and trypanosomiasis. The focused approaches to researching on these diseases is expanding to yield results that will significantly benefit cattle breeders. The development of rinderpest vaccines remaining stable at ordinary temperatures was of high practical value.

Biotechnology in fisheries is recent in Africa and it is being applied essentially in private fisheries sector. Work is mostly confined to hormonal treatments for sex reversal and pawning, detection of fish and shrimps disease and development of fish feed.

INTRODUCTION

In this chapter, biotechnology is defined as "any technique that uses living organisms to make or modify a product, to improve plants or animals, or to develop microorganisms for specific uses" (after the Office of Technology Assessment, US Congress). The document focuses on the development and application of modern biotechnology which is based on the use of new enabling techniques of recombinant-DNA technology, often referred to as "genetic engineering". Biotechnology has great potential to influence and benefit livestock and fisheries. Modern techniques of biotechnology offer the potential of moving any cloned gene from any organism into any other organism and confer much greater precision and speed in achieving results as compared to conventional techniques. In conjunction with conventional technologies, modern biotechnology holds promise of increased and sustained productivity, efficient processing for improved product diversification and utilization, adaptation of product quality to functional requirements, and decreased reliance on agrochemicals and other external inputs.

It may also promote better conservation and use of genetic resources, and environmentally friendly management of natural resources. However, the number of marketable products and their influence at the farm level still seems to be limited, but is likely to increase in the next decade.

Biotechnology also poses certain challenges. These are largely determined by how, where and when it finds application. In general, the fast-paced research, predominantly funded by private-sector investment, and use of intellectual property rights in industrialized countries are seen as evidence that the application of biotechnology will hold the key to competitiveness and comparative advantage in many fields, including agriculture and food.

Biotechnology, with its vast potential and challenges, is thus of the utmost importance to agricultural development, and hence to FAO. The science of developing biotechnological tools cannot make a distinction between developing and developed countries. However, the application of such tools in the development process requires preconditions which are easily found in many developed countries, but which hardly exist in most developing countries. The current paths of research and development have given rise to concern that the disparity in harnessing biotechnologies for agricultural and economic development may increase between industrialized and developing countries.

The preamble to the FAO Constitution requires that the Organization work towards raising levels of nutrition and the standards of living of all peoples, securing improvements in the efficiency of the production and distribution of all food and agricultural products, improving the conditions of rural populations and thus contributing towards an expanding world economy. Therefore, one of the major tasks for the Organization is to ensure that the benefits of biotechnology will be shared by people in the north and the south, in both large and small, and rich and poor countries.

In recent years, various FAO bodies and conferences such as the Tenth Session of FAO's Committee on Agriculture, 1991; the FAO Conference 1989; Sessions of the FAO Commission on Plant Genetic Resources, 1989 and 1991; FAO/CTA International Symposium on Plant Biotechnologies for Developing

Countries, 1989; and the 20th FAO Regional Conference for Asia and the Pacific, 1990, have strongly recommended that FAO, as the lead UN specialized agency for food and agriculture, must respond to the potential and challenges of modern biotechnology for agriculture, forestry and fisheries. This paper, therefore, examines the pattern of biotechnological development in developed and developing countries, the application, impacts and potential of biotechnologies in agriculture, forestry and fisheries. It then analyses some of the issues and concerns in biotechnology such as trade substitution, biosafety, and intellectual property rights, with special reference to developing countries. Finally, it elaborates the elements of FAO's policy and strategies on biotechnology.

BIOTECHNOLOGY IN FORESTRY

Forestry research in Africa, let alone forestry biotechnologies, is not well developed. Some success stories were experienced with rhizobium technology such as coniferous plantations in East and southern Africa and Casuarina plantations in Senegal. However, in recent years, there is a tendency to move towards genetic improvement and in. vitro culture. African forestry research institutions are actively engaged in laying the foundations for genetic improvement research. This is a very necessary prerequisite for biotechnology or any other sophisticated research to be effective.

According to a survey carried out by the FAO Regional Office for Africa (1987), 20 countries in the region, including the North African subregion, had established national forest seed centres, and FAO is helping, through a cooperative programme with France, to establish a seed centre in each Sahelian country. The survey further revealed that about 100 researchers in 22 countries are undertaking some kind of genetic improvement research, but only three countries are engaged in in vitro culture, with considerable inputs from foreign institutions and expertise in two cases. However, most of the research institutions are engaged in research on vegetative propagation for the establishment of clonal plantations.

DEVELOPMENTS IN BIOTECHNOLOGY

There exists a gradient of biotechnologies, depending on the degree of sophistication, complexity, stage of development and application. The lower side of the gradient comprises simpler but widely used techniques such as in vitro culture, rhizobium technology, fermentation. The upper side of the gradient includes advanced techniques involving genetic engineering. The gradient of biotechnologies may be matched with the gradient of national capabilities, economic investment and efforts to provide the possibility of choosing the appropriate techniques and approaches with the most positive impacts. The pattern of use and development of biotechnologies in developed and developing countries thus varies considerably.

THE PATTERN IN DEVELOPED COUNTRIES

In the industrialized countries, a new pattern of biotechnological research funding has emerged. With the availability of property right protection of biotechnologies and prospects of vast markets for biotechnology products and the techniques, the bulk of the research is funded, carried out, and controlled by the private sector. For instance, since 1976 when GENENTECH, the first biotechnology company, was established, the new techniques have spawned a variety of industries that now comprise more than 400 start-up firms, more than 200 established firms that have diversified into biotechnology, and more than 200 supply firms in the United States alone.

The new biotechnology industry in the United States produced pharmaceutical, diagnostic tests and agricultural products worth close to US$2 billion in 1990. A similar trend is seen in Europe and Japan. It is estimated that about 60 percent of the biotechnology research and development funding in industrialized countries is from the private sector. Thus, the private sector has been a major force in enhancing the capability of these countries in this field.

Research institutions in the public sector are now generally required to raise a substantial part of their budget from non-government sources, via contractual research, licensing agreements and royalties. This is tending to increase secrecy over research findings and to hinder free scientific communication. University professors, researchers and government institution scientists are increasingly becoming entrepreneurial and are entering private industry.

Another important trend is that large multinational corporations are purchasing smaller seed and biotechnology companies and diversifying their holdings. This allows them to develop a package sale of chemicals, seeds and equipment.

Heavy involvement of the private sector and market considerations greatly influence the topics and commodities chosen for research. Major crops, commodities and farming systems of great socio-economic importance to the developing world, but of little international market importance, do not figure in the biotechnology research agenda of industrialized countries. Furthermore, these countries are keen to reduce their production costs, increase the productivity, quality and value of their products and, thus, improve their overall competitiveness in the world market.

THE PATTERNS IN DEVELOPING COUNTRIES

Biotechnology facilities are being established in most developing countries. However, the level of research, development and use of biotechnology for agriculture, forestry and fisheries in the developing countries is generally far below the level in the industrialized countries. Among developing countries, the status varies considerably. A few, such as Brazil, Mexico, India, China and The Republic of Korea, have sought to gain full scientific and technological capacity, especially in agricultural biotechnology. Others, such as Indonesia, Malaysia, the Philippines and Thailand and a few countries of Latin America, have built the capacity to apply biotechniques and develop biotechnologies useful for their agriculture and food industries. The participation of the private sector in gaining biotechnological capacity is not significant in most of these countries.

Many developing countries have inadequate funding, poor human resources and limited access to information, resulting in a relatively low-level capacity for research and technology development and exploitation, especially in the field of modern biotechnology research, which is costly and requires highly trained personnel. Most developing countries are vague as to their immediate aims in agricultural biotechnology. Few have appropriate proprietary-right protection systems or mechanisms to increase their access to protected techniques and products. Furthermore, there is negligible involvement of the private sector, which accentuates the problem of insufficient attention to biotechnology.

One of the major constraints on biotechnological development in developing countries is the poor quality and extent of higher education in frontier sciences, especially molecular biology. Besides, there are no or very weak links between universities and research institutions to reinforce each other and to utilize synergistically the scarce trained personnel resources. The university experts play a negligible role in national policy formulation, including that on biotechnology. Furthermore the universities are generally not oriented for efficiency in commercialization and are therefore not able to market biotechnology products of their own or others' inventions.

Some international and donor-country public sector institutions strive to ensure that attention is given to solving, through biotechnology, the food and agriculture problems of developing countries. Among this group are the International Agricultural Research Centres (IARCs) located or operating in developing countries. Capacity exists in several of these centres for utilization of advances in biotechnologies to solve certain problems of plant production and protection and of livestock production and health, and capacity is being built in others.

For example, the International Rice Research Institute (IRRI) is incorporating modern molecular biology in its rice research activities. Its programme is linked closely to the Rockefeller Foundation Biotechnology Network. The World Bank, the International Service for National Agricultural Research (ISNAR) and the Australian Government undertook a major study on the likely impact of modern biotechnology on agriculture. This study is being followed up by the participants by an expansion in the World Bank lending for biotechnology and by ISNAR's establishing an Intermediary Biotechnology Service to provide advisory services to national agricultural research systems.

FUTURE PROSPECTS AND STRATEGIES

Alleviating the widespread food gap and attaining food security or food reliance and nutritional adequacy should be the highest national priority in almost all African countries. In order to achieve this foremost goal, production, productivity and sustainability of food and agriculture must be greatly enhanced through the generation, development and transfer of appropriate technologies. While the potential of currently available technologies should be rationally harnessed, new and emerging technologies such as biotechnologies, should be used in conjunction with, not as substitute for, existing conventional technologies.

Biotechnology is offering innovative possibilities for increasing crops, livestock, forestry and fisheries production and for protecting the environment by the reduced use of agrochemicals, conservation of genetic resources, bioremediation and recycling of wastes. In vitro culture techniques for micropropagation, freeing planting materials of viruses and other pathogens, the production of secondary metabolites and germplasm conservation are being widely used.

Genetic mapping of major tropical crops as an aid to conventional plant breeding programmes; plant virus resistance by genetic engineering of the host plant; novel biocontrol agents for pest control to reduce pesticide use; and the use of efficient and new biofertilizing agents are now being increasingly used mostly in developed and in a few developing countries. Embryo transfer techniques and the use of diagnostic kits for detection of diseases both in plants and animals, and the use of highly specific and effective vaccines are proving of great value to animal production and health.

Biotechnologically designed new feeds, hormones and disease control agents are already favourably impacting fish production. But, in Africa, these technologies and products are being used only rarely, in a limited way and in an ad hoc manner. The slow and poor development of modern biotechnology in Africa may be ascribed to: (i) lack of appropriate policies, strategies, plans and programmes and the absence of political will in general; (ii) inadequacy of human resources, paucity of financial investments, poor research and laboratory infrastructure, absence of priority-setting mechanisms and insufficiency of mechanisms to analyse global market and technology trends; and (iii) negligible involvement of the private sector and absence of effective regulatory measures.

In view of the above, the African strategy for biotechnological development should include both the provision of direct support for biotechnological research, product development, technology assessment and transfer and the creation and fostering of a climate where biotechnology can develop in a balanced manner. This will largely depend on governmental and public awareness, political will, policies and programmes for building national capabilities and strengthening/developing regional and international cooperation.

Creative partnerships between public and private sector interests are critical in establishing a competitive strategy in biotechnology. While developing institutional mechanisms best suited to a given country, it is essential to establish effective interdisciplinary collaboration among molecular biologists, breeders, agronomists and plant protection specialists and to forge and menage linkages among researchers, extensionists, farmers and other users of bioproducts and techniques. It may be generally preferable to build the new initiatives on traditional strengths in agricultural, forestry and fishery research, rather than creating totally new institutes. It should be kept in mind that biotechnology in crop and livestock production will be applied through the creation of new genotypes possessing novel characteristics, and this is possible only when conventional breeding and production systems are already in place.

Development of trained personnel should be a high priority in Africa. The important areas of training that are required for a successful biotechnology programme are microbiology, plant biology, zoology, cell biology, virology, genetics, biochemistry and cell and molecular biology. In most of sub-Saharan Africa, the people trained in these areas are in very short supply. Even biology and tissue culture training is not in the currriculum of most tertiary education institutions. Very few universities offer training with adequate laboratory facilities in molecular biology and recombinant DNA technology.

Those that are trying to offer such training are constrained by the lack of equipment and staff. In recent years, however, some donor-supported formal training programmes in biotechnology have been developed. For instance, the M.Sc. in biotechnology at the University of Zimbabwe was assisted by SAREC (Sweden) and DGIS/Ministry of Foreign Affairs (the Netherlands).

The donors provided the funding for both the teaching/research equipment as well as funds to buy supplies and to cover the cost of bringing Dutch and Swedish biotechnology specialists to Zimbabwe to give special lectures in areas in which there was no local competence. Students from Zimbabwe have also been admitted to universities in the donor countries for Ph.D. degrees. Such arrangements exist and/or are being developed in several other African countries. The actual and potential effects of biotechnology development are not always positive. In the African context, the negative effect is the displacement of crops at present grown for export in Africa by biotechnologically designed "industrially" produced analogous in industrialized countries.

African countries should be aware of such developments and diversify their products or increase quality and cost-effectiveness of their products not only to maintain the international market share but also to enhance or open new opportunities. FAO and other concerned international systems should assist African countries in developing early warning systems whereby the potential relative substitution effects could be monitored and strategic adjustments brought about in time to avert damage.

In industrialized countries, where biotechnology is essentially in the hands of the private sector, the work is focused on areas and commodities that will fetch higher economic returns to often costly investments in biotechnology research and development. Commodities such as bananas plantations, cocoa, oil palm, roots and tubers, which are of high socio-economic importance to Africa, should be on priority biotechnology research agendas of countries in the region. International agencies and research centres should also pay due attention to "orphan" commodities.

Most of the African countries with an interest in modern biotechnology are aware of the need to institute adequate biosafety provisions and establish appropriate intellectual property and patenting systems. The GATT agreements have brought these issues to the freefront, and they must be attended to in time to enable countries to access global markets. Preliminary work on developing biosafety guidelines is underway in many African countries.

The status of the issue of intellectual property rights in the region is still in a state of flux. Policymakers and biotechnologists in the region, in consultation with concerned international agencies, need to address this issue more decisively. Access to information by establishing the necessary national and regional databases and information systems should be strengthened to render priority setting, technology assessment and transfer mechanisms more effective.

BIOTECHNOLOGIES FOR FISHERIES IN DEVELOPING COUNTRIES

Biotechnology in fisheries and aquaculture represents a range of technologies that present opportunities to increase growth rate in farmed species, to improve nutrition of feeds for aquaculture, to improve fish health, to help restore and protect environments, to extend the range of aquatic species and to improve management and conservation of wild stocks. In this e-mail conference, the focus will be on genetic biotechnologies, with a brief treatment of related reproductive and gene banking technologies, and the appropriateness of their application in the fishery sector in developing countries. It is important to note that developing countries produce more fishery products from aquaculture, inland capture fisheries and marine capture fisheries than developed countries. The coverage of the biotechnologies here is not comprehensive, but should be enough to stimulate discussion in the conference.

The vast majority of aquatic genetic resources are found in wild populations of fishes, invertebrates and aquatic plants. Fishstat, the FAO database on fishery statistics, lists 1,235 taxa of common aquatic species that are harvested by humans in major fisheries; thousands more species are taken by small-scale fishers. It also contains information on 440 species that are farmed, but just 20 of these taxa account for approximately 80% of world aquaculture production. Domestication of aquatic species has not proceeded to the same level as it has in the crop and livestock sectors. Genetic biotechnologies must be used both to assist in the further domestication of aquatic species and to help manage and conserve the genetic resources found in wild populations.

GENETIC BIOTECHNOLOGIES IN THE FISHERY SECTOR

This Background Document provides a summary of recently developed biotechnologies that could be used, or more widely used, in the fishery sector in developing countries. Genetic biotechnologies that can be used in fisheries and in aquaculture include those that help to manage genetic resources and those for genetic improvement. For management of genetic resources, markers can be used in the identification of management units and of endangered species to assist fishery management and they can also help broodstock management in stocking programmes. These markers may be genes, proteins (i.e. the products of genes), sequences of DNA or the phenotypic expression of genes (different colours, shapes etc.). In the 1960's, analysis of proteins revealed a wealth of genetic diversity in wild populations. Protein analysis is now relatively fast and inexpensive, but it requires tissue samples to be stored and transported frozen. DNA analysis is becoming the method of choice because of the small amount of tissue needed, the fact that the tissue can be stored dried or in alcohol, and because DNA analysis reveals much more genetic variation than protein analysis.

Several kinds of DNA markers exist, such as RFLPs, AFLPs, RAPDs and microsatellites. These, as well as other kinds of markers, can be used to analyse gene frequencies and genetic variation in and between different groups of fish. Studies carried out using these technologies in fish populations have revealed high levels of genetic variation distributed throughout the fish genome.

Genetic improvement technologies cover a range of techniques requiring different levels of expertise and resources. Chromosome-set manipulation (i.e. polyploidy induction) is an established technique to increase the number of chromosome-sets (ploidy number) in an organism. Temperature, chemical and pressure shocks applied to fish eggs can be used to produce triploid (3 chromosome-sets) individuals that have desirable culture traits. Sex-reversal and the production of single sex groups of fish is also a simple technology that combines hormone treatment and chromosome-set manipulation.

Hybridisation, i.e. the mating of genetically different groups from the same species (intra-specific hybridisation) or from different species (inter-specific hybridisation), is a simple technique that is now easy to accomplish due to our increased knowledge of reproductive biology. It can be used to combine good traits from two different species into one group of fish or to transfer a characteristic of one group to another. A problem is that breeding hybrids with hybrids results in a non-uniform and unpredictable group of fish that is generally not well suited for culture. Therefore, for hybrid production, the parent-lines must be maintained pure. The above genetic improvement techniques are considered short-term strategies, where the gains are seen in one or two generations. Selective breeding is a longer-term strategy where gains are accumulated at each generation of selection. Molecular markers may now increase the efficiency of selective breeding by facilitating the identification of quantitative trait loci (QTLs), i.e. genes that control complex characters such as growth rate and environmental tolerance and, secondly, by making it possible to use molecular markers linked to QTLs to identify desirable individuals or families.

Genetic engineering and the production of transgenic organisms is an active area of research and development in aquaculture. This is a medium-long term strategy in that development and testing of stable transgenic lines requires time. The large size and hardy nature of many fish eggs allows them to be manipulated rather easily and facilitates gene transfer by direct injection of a foreign gene or by electroporation, where an electric field assists gene transfer.

In the next three sections, we will briefly discuss currently available biotechnologies in the context of fishery management, aquaculture and conservation respectively.

Bibliography

Allen Pickham: *Salmon and His People: Fish and Fishing in Nez Perce Culture,* Lewiston, Idaho: Confluence Press, 1999.

Arnold C. Long: *Deep Sea Demersal Fish and Fisheries*, Cyber Tech Pub, Delhi, 2009.

Bailey, Ronald: *Liberation Biology: The Scientific and Moral Case for the Biotech Revolution*, Amherst, Prometheus Books, 2005.

Biswas, K.P. : *Economics in Commercial Fisheries,* Daya, New Delhi, 2006.

Boyd, Robert: *People of The Dalles: The Indians of Wascopam Mission*, Lincoln, University of Nebraska Press, 1996.

Charles, A. T.: *Sustainable Fisheries Systems,* Oxford: Fishing News Books, Blackwell Scientific, 2001.

Chrispeels, Maarten : *Plants, Genes and Crop Biotechnology,* Sudbury MA, Jones and Barlett Publishers, 2003.

Clarence Woodcock: *In the Name of the Salish and Kootenai Nation: The 1855 Hell Gate Treaty and the Origin of the Flathead Indian Reservation*, Pablo, Montana: Salish Kootenai College Press, 1996.

Cohen, Fay: *Treaties on Trial: The Continuing Controversy over Northwest Indian Fishing Rights,* Seattle. University of Washington Press, 1986.

Cone, Joseph: *A Common Fate: Endangered Salmon and the People of the Pacific Northwest,* Corvallis, Oregon State University Press, 1995.

Curtis, H.: *Introduction of Embryology of Fisheries,* New York, Worth Publishers, 1989.

Dholakia, A.D. : *Fisheries and Aquatic Resources of India,* Daya, Delhi, 2004.

Drexler, K. Eric: *Engines of Creation: The Coming Era of Nanotechnology,* New York, Anchor Books, 1986.

Edward A.: *Papers on Bacterial Genetics,* Boston, Brown and Company, 1960.

Henry I. : *The Frankenfood Myth: How Protest and Politics Threaten the Biotech Revolution,* New York, Praeger, 2004.

John H.: *Fish Technology,* New York, Cambridge University Press, 1985.

Kelleher, K. and P. Cacaud: *Governance of Yemen's Fisheries: Analysis and Project Financing Proposal,* Prepared for the European Commission, Tebodin, 1999.

Mahadevan, R.K. : *Marine Capture Fisheries,* Campus Books, Delhi, 2008.

Melissa Hirsch Kuchma: *Fisheries Law Enforcement: Assessment of Deterrence,* Institute for Defense Analyses, Virginia, 1999.

Metting, Jr., F.B.: *Soil Microbial Ecology: Applications in Agricultural and Environmental Management,* Marcel Dekker, New York, 1992.

Murray, David: *Seeds of Concern: The Genetic Manipulation of Plants*, Sydney, University of New South Wales, 2003.

Omprakash Sharma: *Handbook of Fisheries and Aquaculture*, Agrotech Publishing Academy, New Delhi, 2009.

Peterson, Keith: *River of Life, Channel of Death: Fish and Dams on the Lower Snake*, Lewiston, Idaho: Confluence Press, 1995.

Primrose, S.B.: *Principles of Gene Manipulation*, London, Blackwell Scientific Publications, 1989.

Ramakrishnan Korakandy: *Fisheries Development in India : The Political Economy of Unsustainable Development*, Kalpaz Pub, Delhi, 2008.

Richards, A.: *Reported Fisheries Violations in the EEZs of FFA Member Countries, 1978-2001: An Analysis of the Violations and Prosecutions Database*, Forum Fisheries Agency, 2001.

Rodney A.: *Flesh and Machines: How Robots Will Change Us*, New York, Pantheon Books, 2002.

Roobeek A: *The Fish Biotechnology Revolution?*, Oxford, Blackwell, 1995.

Sakhare, V B : *Reservoir Fisheries and Limnology*, Narendra Pub, Delhi, 2007.

Srivastava, C.B.L. : *A Text Book of Fishery Science and Indian Fisheries*, Kitab Mahal, Delhi, 1999.

Steve Warner, Melissa Hirsch Kuchma: *Fisheries Law Enforcement: Assessment of Deterrence*, Institute for Defense Analyses, Virginia, 1999.

Taylor, Joseph: *Making Salmon: An Environmental History of the Northwest Fisheries Crisis*, Seattle, University of Washington Press, 1999.

Tomi Petr: *Fisheries in Irrigation Systems of Arid Asia*, Daya, New Delhi, 2007.

Ulrich, Roberta: *Empty Nets: Indians, Dams, and the Columbia River*, Corvallis, Oregon State University Press, 1999.

Umesh C Goswami: *Natural and Anthropogenic Hazards on Fish and Fisheries*, Narendra Pub House, Delhi, 2007.

Walden, Richard: *Genetic Transformation in Plants*, England, Open University Press, 1988.

Index

M

N

P

R

S

W